MAPPING THE FUTURE OF BIOLOGY

BOSTON STUDIES IN THE PHILOSOPHY OF SCIENCE

For further volumes:
http://www.springer.com/series/5710

MAPPING THE FUTURE OF BIOLOGY

EVOLVING CONCEPTS AND THEORIES

Editors

ANOUK BARBEROUSSE
MICHEL MORANGE
THOMAS PRADEU

Springer

Editors
Anouk Barberousse
Université Paris
Panthéon-Sorbonne
Paris, France

Michel Morange
Université Paris
Panthéon-Sorbonne
Paris, France

Thomas Pradeu
Université Paris
Panthéon-Sorbonne
Paris, France

ISBN 978-90-481-8175-9 e-ISBN 978-1-4020-9636-5

DOI 10.1007/978-1-4020-9636-5

Cover design: Boekhorst Design BV

Printed on acid-free paper

9 8 7 6 5 4 3 2 1

springer.com

Foreword

Carving Nature at its Joints?

In order to map the future of biology we need to understand where we are and how we got there. Present day biology is the realization of the famous metaphor of the organism as a *bête machine* elaborated by Descartes in Part V of the *Discours*, a realization far beyond what anyone in the seventeenth century could have imagined. Until the middle of the nineteenth century that machine was an articulated collection of macroscopic parts, a system of gears and levers moving gasses, solids, and liquids, and causing some parts of the machine to move in response to the force produced by others. Then, in the nineteenth century, two divergent changes occurred in the level at which the living machine came to be investigated.

First, with the rise of chemistry and the particulate view of the composition of matter, the forces on macroscopic machine came to be understood as the manifestation of molecular events, and functional biology became a study of molecular interactions. That is, the machine ceased to be a clock or a water pump and became an articulated network of chemical reactions. Until the first third of the twentieth century this chemical view of life, as reflected in the development of classical biochemistry treated the chemistry of biological molecules in much the same way as for any organic chemical reaction, with reaction rates and side products that were the consequence of statistical properties of the concentrations of reactants. But then, in the middle of the century, biochemistry gave way to molecular biology in which the three-dimensional structure of individual macromolecules rather than the laws of mass action became the basic element in prediction and understanding. Molecular biology is not simply the biochemistry of a certain class of biologically important compounds. It is a reformation of explanatory narratives in which the individual molecule and its detailed three-dimensional structure are at the center of explanation. The description of the chain of transformations of these structures is, in a sense, a return to the form of the machine narrative exemplified by Descartes' and Harvey's description of the circulation of the blood. It is easy to predict, then, that in the near future the three dimensional elucidation at the Angstrom level of resolution of biological molecules and of the interactions between those molecules, will continue to increase in importance and will, for some period, come to dominate functional biology. No explanation of cellular and organismic function including the study of development will be considered complete without a description of the three-dimensional writhings, couplings and metamorphoses of individual macromolecules. But what then?

Living organisms are not simply assemblages of molecules. They differ from inanimate assemblages in three respects. They are open, internally functionally heterogeneous, complex systems. Organisms are in a constant state of interchange with the world outside. Unlike inorganic structures, their persistence depends upon a constant intake and export of molecules. Their internal molecular structures are hierarchically organized both spatially and functionally into subsystems and subsystems of subsystems that perform various operations of maintenance, growth, reproduction, and metabolism. But these subsystems are not independent in their functions, feeding back on one another. We do not have a nervous system *and* an endocrine system, but a neurosecretory system in which a change in the secretory activity of the endocrines is both a cause and an effect of mental states. Moreover, what is not widely appreciated is that the number of copies of biologically important molecules within cells is very low so that chemical transformations cannot depend on the laws of mass action. Individual molecules must be at the right place at the right time in the cell, and in the right vibrational state, in order to carry out essential metabolic activities. In the most extreme case, that of DNA in a haploid cell, there is only a single molecular copy, yet the entire future of that cell and its progeny depend upon the successful completion of DNA replication before cell division and the guaranteed apportionment of one and only one daughter molecule to each daughter cell. The molecular events and structures of cell division cannot be understood without taking account of this problem.

Thus, the program of biology cannot be realized simply by a knowledge of the structure of individual molecules and how those structures enter into individual reaction sequences. Biology is essentially a study of anatomy. For Harvey and Descartes and their Greek precursors that meant cutting up the body into individual organs of definable function. But even that level of analysis was seen as destructive. Alexander Pope warned that.

> "Like life following life in creatures you dissect, You lose it in the moment you detect."

For modern molecular biology it is the structure of the individual molecule that counts. But the program of biology must be to connect these two extremes of structure by the study of the hierarchical organization of structures of the cell and body. There is already a long-standing consciousness of this program. The study of the structure of membranes and of cell organelles is now common and an issue of *Nature* or *Science* is likely to contain an article showing both the structure of an individual macromolecule and the structural assemblage of macromolecules making up membrane pores or cell organelles. This is the next step in what eventually must be the modal pattern for the level-by-level study of biological structure in order to elucidate function.

There remains one nagging doubt. Will biology ever be able to elucidate the relation between particular mental states and physical states of the nervous system? What is the spatio-temporal pattern of nerve impulses in the brain that had as its mental manifestation the framing of the thought, "Today is Thursday"? And if precisely the same sentence were constructed next week or by a different person would the pattern of nerve states be the same? Attempts to correlate brain states with mental states by brain imaging techniques that detect either electrical or circulatory activity are too crude to provide neural pathway information. What the standard explicatory model of biology demands in the case of mental activity is that given some physical description of the firing of neurons, the mental state of consciousness, " Today is Thursday" (in English), can be unambiguously inferred. We do not know how to proceed, even in principle, to satisfy such a demand. We may never know.

Richard C. Lewontin

Contents

Contributors

Anouk Barberousse Institut d'Histoire et de Philosophie des Sciences et des Techniques, CNRS, Paris, France, anouk.barberousse@ens.fr

Mark A. Bedau Professor of Philosophy and Humanities, Reed College, Portland, OR, USA, mab@reed.edu

Scott F. Gilbert Howard A. Schneiderman Professor of Biology, Swarthmore College, Swarthmore, PA, USA, sgilber1@swarthmore.edu

Peter Godfrey-Smith Professor of Philosophy, Harvard University, Cambridge, MA, USA, pgs@fas.harvard.edu

Richard Lewontin Alexander Agassiz Professor of Zoology and Biology, Harvard University, Cambridge, MA, USA, lewontin@oeb.harvard.edu

Evelyn Fox Keller Professor of the History and Philosophy of Science, MIT, Cambridge, MA, USA, efkeller@mit.edu

Michel Morange Professor of Biology, Ecole normale supérieure (ENS) and Paris 6 University, Paris, France, morange@wotan.ens.fr

John Odling-Smee Institute of Biological Anthropology, University of Oxford, Oxford, UK; Director of Studies & Lecturer in Human Sciences – Mansfield College, Oxford, UK, john.odling-smee@anthro.ox.ac.uk

Susan Oyama Professor Emeritus, Psychology Department, John Jay College, New York, NY, USA, soyama48@verizon.net

Thomas Pradeu Assistant Professor, Paris-Sorbonne University, Paris, France, thomas.pradeu@paris-sorbonne.fr

Michael Ruse Professor, Department of Philosophy, Florida State University, FL, USA, mruse@mailer.fsu.edu

Kim Sterelny Professor, Philosophy Programme, RSSS, Australian National University, Canberra, Australia, kimbo@postbox.anu.edu.au

Chapter 1
Introduction

Anouk Barberousse, Michel Morange and Thomas Pradeu

1.1 Molecular Biology Meets Evolutionary Biology: Challenges to Post-Mayrian Biology

1.1.1 The Mayrian Gap

In one of his most famous papers, Ernst Mayr (1961) claims that the modes of explanation of molecular biology and evolutionary biology are completely different, because the former refers to *proximate causes*, whereas the latter resorts to *ultimate causes*. At the celebration of his 100th birthday, only a few months before his death, he asserted that molecular biology of the second half of the 20th century, despite spectacular progress, had not had any real impact on the synthetic theory of evolution:

> By the end of the 1940s the work of the evolutionists was considered to be largely completed, as indicated by the robustness of the Evolutionary Synthesis. But in the ensuing decades, all sorts of things happened that might have had a major impact on the Darwinian paradigm. [. . .] Unexpectedly, however, none of these molecular findings necessitated a revision of the Darwinian paradigm—nor did the even more drastic genomic revolution that has permitted the analysis of genes down to the last base pair.
>
> It would seem justified to assert that, so far, no revision of the Darwinian paradigm has become necessary as a consequence of the spectacular discoveries of molecular biology. (Mayr 2004).

We believe that only few biologists or historians/philosophers of biology currently share Mayr's conviction that evolutionary and molecular biology have had no real influence one on the other for the past sixty years.[1] As early as the 1960s and 70s, many biologists emphasized how new findings in molecular biology influenced

---

A. Barberousse (✉)
Institut d'Histoire et de Philosophie des Sciences et des Techniques, CNRS, Paris, France
e-mail: anouk.barberousse@ens.fr

[1] Of course, if Mayr had been referring to the 'core' of evolutionary theory – say the three necessary and sufficient conditions for evolution by natural selection given by Lewontin (1970) (heredity, variation, differential fitness) – he could have said that the "Darwinian paradigm" remained untouched, but it is definitely not what the 2004 quotation is about, since it deals with the synthetic theory of evolution.

A. Barberousse et al. (eds.), *Mapping the Future of Biology*, Boston Studies in the Philosophy of Science 266, DOI 10.1007/978-1-4020-9636-5_1,

evolutionary biology. For instance, the discovery of genetic drift as a major component of evolution testifies the importance of such an influence: in this case, molecular and evolutionary biology are closely articulated (Kimura 1968). Moreover, it was discovered through molecular investigations that mutations are much more numerous than it was expected, and that most of them are not submitted to natural selection. Another example is the discovery of molecular homologies and the realization of their evolutionary effects (Britten and Kohne 1968; analyzed by Gilbert, Opitz and Raff 1996). Other examples would include those given by Crick (1966), Jacob (1977) or Gould (1977). The influence of molecular biology on evolutionary biology is also well documented by the work of historians and philosophers of biology (Judson 1979; Keller and Lloyd 1992; Waters 1994; Chadarevian and Kamminga, eds., 1998; Gayon 1998; Morange 1998; Gilbert 2000; Amundson 2005). Today, the great majority of biologists and philosophers of biology consider that evolutionary biology and molecular biology (broadly defined as the set of disciplines looking for molecular explanations and using molecular tools like DNA sequencing and PCR: developmental biology, molecular genetics, genomics, immunology, molecular neurology, etc.) can and must be articulated (Gilbert et al. 1996; Burian 1997; Griffiths 2002; Morange 2005; Weber 2005).

The difficulties of articulation have often been neglected; this book focuses on them. The contributions gathered here, written by some of the most influential theoretical biologists and philosophers of biology of the last twenty years, all consider that the articulation of evolutionary and molecular biology raises important issues, first and foremost that of what can be considered as a valid explanation in today's life sciences. They take seriously the growing emphasis on the need to connect and even in some cases to unify, the modes of explanation of molecular biology and those of evolutionary biology. As far as we know, this book is the first attempt to show what makes this articulation between different modes of explanation desirable, how it can be done, and, even more importantly, how it challenges today's theoretical biology.

When trying to articulate molecular and evolutionary approaches to find out the solution of a given problem, several types of difficulties may arise. Some are of a general theoretical nature, for instance the question of whether evolution by natural selection is constrained by principles of self-organization. Others are conceptual: some concepts play a key role in today's biology but are in need of a precise and rigorous definition. A first example is the gene concept: molecular and evolutionary biologists seldom use the same concept of a gene. Another example is the organism concept, a sort of blind spot in today's biology, since only few explicit definitions of this concept are available. In the next subsection, we review three domains of molecular biology in which difficulties of both types are experienced and we indicate how the contributed papers relate to these "awkward" domains. In Section 1.2, we show that contrary to what we were expecting before this volume was planned, it is when they try to solve the difficulties of the *second* type in each relevant domain that biologists and philosophers do invent future biology: the papers gathered here testify that research generated by the exploration of some well-chosen concepts is probably more fruitful than attempts to erect brand new theoretical alternatives to neo-Darwinism.

1.1.2 Some Awkward Domains

(*i*) *Genetics: The definition of the gene and the metaphor of the genetic programme:* The 20th century may well have been "the century of the gene" (Keller 2000). However, the great advancements of genetics have been achieved without a clear definition of what a gene is. When the structure of DNA was discovered, many believed that molecular biologists had found the material substance corresponding to the abstract entities postulated since the rediscovery of Mendel's laws and that Mendelian genetics could be rapidly and entirely reduced to molecular genetics. But it became more and more obvious that whereas this reduction was unproblematic from the practitioners' point of view, it was conceptually more difficult (Hull 1974; Rosenberg 1985; Waters 1990). This inter-theoretical reduction raised questions about the relationship between the molecular machinery of reproduction and the structure of inheritance. The concept of a gene is central to both; however, the word "gene" has many different and sometimes conflicting meanings in today's biology (Falk 2000; Rheinberger and Müller-Wille 2004; Morange, this volume; Oyama, this volume; Godfrey-Smith, this volume). At least *two* concepts of the gene were involved, that may be respectively characterized as the "molecular" and the "evolutionary" concepts, and it appeared that articulating them was a very complicated task (Kitcher 1984; Maienschein 1992; Waters 1994). The *evolutionary* gene is an abstract, functional entity: it is a unit of heredity explaining the inheritance of phenotypic traits. The only important thing is that it is associated with a heritable phenotypic difference on which natural selection can act. The *molecular* gene is a fragment of DNA which has a well-identified function. The two "genes" do not always coincide, due in particular to pleiotropy and epistasis. Lenny Moss (2003) makes a similar distinction between "genes-p" (genes defined via the phenotype they are associated with) and "genes-d" (fragments of DNA identified by their molecular role in development) (see also Griffiths and Neumann-Held 1999; Stotz and Griffiths 2004). All these works reveal an urgent need for conceptual clarification. Because of the many meanings of the concept of gene and the confusions they allow, some people today doubt that it is still useful. They suggest to using different terms for the various meanings (Griffiths 2007; see also Pearson 2006).

Another, but related, controversy, concerns the notion of a *genetic programme*, which has been challenged in various ways (Lewontin et al. 1984; Nijhout 1990; Oyama 2000b [1985]; Keller 2000; Moss 2003; Oyama, this volume). Some biologists consider that this notion is too outdated to merit criticism. But they should not forget that it was biologists who abundantly used this notion in the first place, and sometimes promoted its use (among many others: Jacob 1973; Monod 1970). Moreover, the notion of a genetic programme is still widely used nowadays in cognitive science, developmental biology, systems biology, etc.

One obvious consequence of these challenges is that the role of genes in inheritance is now seen differently than it had been. Recently, genes have lost a part of their previous explanatory power (Moss 2003) and their role has proved to be much more context-dependent than it was thought (Griffiths 2007). As early as the 1980s, numerous proposals were made by biologists and philosophers about the role of genes in inheritance (e.g. Lewontin 1983a). Recently, radical suggestions have been formulated. One of them is to adopt global units of analysis for the understanding of inheritance instead of genes, in particular "developmental systems" (Oyama 2001). Simply abandoning the phrase "genetic programme" may well not be enough, because some biologists who do not use this concept anymore are still prone to similarly confusing statements, particularly when they speak about "information" (Oyama 2000b [1985]; Oyama this volume). Susan Oyama shows here that some conceptions of biological information carry the idea of a pre-existing form which would be the *maker* of the organism and would explain heredity. Together with Peter Godfrey-Smith (this volume), they emphasize the danger of reifying the notion of information (see also Sarkar 1996a; Godfrey-Smith 1999, 2000a; Griffiths 2001).

Another important consequence of these discussions is that the "gene's eye view" of evolution (Williams 1966; Dawkins 1976), which is largely dominant among evolutionary biologists (who commonly define evolution as change in gene frequency) currently undergoes significant challenges. The debate between genocentric and organism-centred or multi-levelled views of evolution is still unresolved. It will probably have a major impact on how we conceive of evolution and on the relationships between evolutionary and molecular biology.

(*ii*) *Articulating development and evolution:* It is generally admitted that the synthetic theory of evolution did not integrate developmental biology. A few decades ago, strong claims were made about the arising of the long-awaited "evo-devo" synthesis, namely the unification of evolutionary theory and developmental biology. Darwin himself considered developmental studies as an important place where the principles of the theory of evolution should be validated. According to Goldschmidt (1940) and Waddington (1947), about a century later, all changes important in evolution are alterations in ontogenesis, or development. However, evolutionary studies, and especially population genetics, have been artificially separated from developmental studies for about a century, during which development was regarded as a "black box" the uncertain products of which would hardly affect the work of evolutionary biologists. According to Buss (1987), this conception of the relationship between evolutionary biology and developmental biology partly originated in Weismann's work, but is still present today (Dawkins 1982).

The situation started to change when Gould and Eldredge published their work on "punctuated equilibria" (Eldredge and Gould 1972; Gould and Eldredge 1977). They insisted on the importance of developmental constraints, which lead to a limitation of the phenotypes upon which natural selection may act (see also: Gould and Lewontin 1979, discussed by Ruse in this volume; Alberch 1982; Raff and Kauffman 1983). Gould (1977) attempted to further articulate ontogeny and phylogeny – despite what he described as the "unpopularity" of such a project.

The evo-devo synthesis *per se* emerged in the 1980s. It was expected to put an end to the abnormal division between developmental and evolutionary studies. At the beginning of the evo-devo approach, the main questions were:

(i) How does natural selection act on ontogenetic processes?
(ii) To what extent do developmental processes introduce constraints onto evolutionary change?

The discovery and detailed study of homeotic genes, whose number and range of expression have been only slightly transformed since the Cambrian Revolution, has provided a vast amount of knowledge about macroevolution, and has given rise to the invention of "models of development" trying to account for these new findings (Wimsatt 1986; Wimsatt and Shank 1988; Rasmussen 1987). Wimsatt's notion of "Generative Entrenchment" (Wimsatt 2001 for a review), for instance, is meant to explain what kinds of constraints developing systems put on natural selection. The discovery of homeotic genes has been followed by the formulation of many hypotheses, including the "developmental revolution" hypothesis according to which transformations of animal's body plans are generated by changes in developmental processes (Davidson 1995, 2001; Peterson and Davidson 2000; Erwin and Davidson 2002). Two interpretations of the Cambrian Revolution have thus emerged, one focused on developmental *constraints*, giving rise to an approach of evo-devo that is more grounded in embryology, and the other on the role of the environment.

These controversies remain vivid (Robert 2004). The debate over the role of developmental constraints on evolution is still going on (Gilbert et al. 1996; Ruse, this volume; see also the debate stemming from the Paris conference "The Making up of Organisms" between Michael Ruse and Scott Gilbert: Ruse 2006a,c; Gilbert 2006a). Questions about model organisms are arising, and may influence the view we have of development. Indeed, the most widely-used instruments in the investigation of evo-devo are model organisms. While first used in genetic studies, model organisms are now viewed as susceptible of answering new questions. Correspondingly, new models have appeared, far beyond the classic flies, mice and yeast. Gilbert (this volume) detects an important change in our conception of the utility of model organisms in the fact that environment (vs. test tubes) is more and more present. He then draws the consequences of these changes on our conception of how to model development. Finally, another challenge to a complete evo-devo synthesis has been recently emphasized: many, if not most multicellular organisms share (part of) their life span with symbionts which play sometimes necessary physiological roles within them. How is a conception of development centred on the primary action of genes to account for this widespread fact? Does not the ubiquity of symbionts force us to a major reconceptutalization of development as well as of evolution? (McFall-Ngai 2002; Sterelny 2004; Gilbert 2006b).

How far has evo-devo come today? Developmental studies now take up a central place in evolutionary biology, owing to the variety of data they have recently produced and the theoretical fecundity they have gained from experimental work

(Buss 1987; Amundson 2005; Gilbert 2006a). Does this imply that evolutionary biology is about to experience a "major reorientation" because of the influence of developmental biology (Raff 1996)? How should developmental biology be articulated to evolutionary biology? The need for such an articulation, that will perhaps change our conception of evolution (Gilbert 2006a; Sterelny this volume), is beginning to make itself felt in a most urgent way. Trying to determine what exactly the consequences of the influence of developmental biology on evolutionary biology will be, and the other way round, is probably one of the most exciting challenges faced by today's biology and philosophy of biology (Ruse 2006a). The urge to build an even bigger picture, integrating ecology, evolution and development or eco-evo-devo (Gilbert 2001; Gilbert, this volume; Odling-Smee, this volume) is still more challenging.

(*iii*) *Self-organization and complexity:* Evolutionary biology has recently experienced a major challenge coming from the frontier of physics and biology: it has been claimed from different parts that there are laws of complex systems, including living ones, that severely constrain the mechanisms of evolution (Kauffman 1993; Depew and Weber 1995; discussed by Evelyn Fox Keller in this volume). According to this vision of evolutionary dynamics, natural selection does not act in an open space of phenotypic possibilities; its role is rather limited in advance by the emerging physical properties supposedly revealed by Kauffman's models. Some even argue to the effect that there are laws of (living) complex systems directing evolution by natural selection (Bedau, this volume; McShea 2005). More generally, the words "complexity" and "self-organization" have never been so widely used in biological papers or in leaflets presenting biological institutions, which indicates that these themes are strongly attractive nowadays.

Kauffman (1993) claims that physico-chemical processes act as constraints on evolution, and produce "order for free", that is, order not obtained by natural selection. According to this view, natural selection is an important designer of adaptations, but not the only one: its action is pre-determined by the laws of self-organizing complex systems (discussed by Ruse, this volume). Part of the research on self-organizing systems is applied to organic molecules. This suggests that molecular biology, at least when it is done with an eye on self-organizing processes, may challenge our usual views on evolution by natural selection. Within the domain of biological self-organization and complexity, the issue of connecting molecular and evolutionary biology is supposedly solved by postulating principles filling the "Mayrian gap", namely principles of self-organization of autocatalytic systems or general trends of complexification within the living world. More generally, Kauffman's influential works have pressed many to reject, at least in their statements of principle, the neo-Darwinian orthodoxy as it is based on a strong emphasis on the action of natural selection. The role of natural selection is lessened in this new theoretical perspective.

A more and more developed subfield in the study of biological complexity and self-organization is the emergence of life. It was neglected by the first molecular biologists, but it is now treated with renewed interest, and has become the subject of various controversies (Hazen 2005; Luisi 2006). One important question concerns the possibility of life (or proto-life) *before* the appearance of natural selection: is it possible to extend back the action of natural selection to the pre-biotic era, or did the emergence of life reflect a phase transition in progressively complexifying and self-organizing pre-biotic systems (Keller, this volume)? Some (e.g. Kitano 2004) suggest that evolution by "pre-Darwinian" selection may be characterized by self-organization and robustness principles. If natural selection arose long after the appearance of life, is it possible to say *when* this event occurred? And, perhaps even more importantly, what do laws of self-organization and robustness tell us about the definition of life? To define life, one usually resorts to three characteristics: variation, reproduction and natural selection; however, if life appeared before natural selection, how should we define it?

Our view about this domain is that its promise of unifying biological schemes of explanation by encompassing them in a more physico-chemical framework still has to be made convincing. As one of us argues (Morange, this volume), no empirical research programme has yet emerged that could show that betting on self-organization is a fruitful approach.

When planning to organize Paris 2006 workshop on the role of models and theories in today's biology, our impression was that the most important issues were coming from new general *theoretical* approaches, namely that the theories of self-organization, niche construction or epigenetics, were to be seen as *alternative approaches* to neo-Darwinism. However, the conference proved that neo-Darwinism is still the only unifying theoretical approach to the living world and that the newer trends have to be conceived of as *complementing* neo-Darwinism, rather than as competing with it.

Novelty in today's biology is coming from the realization that some concepts which are widely used in different or even disconnected contexts, but are still vague and in need of clarification, may yet be given reinforced methodological roles when clarified. Examples include the concepts of multi-realizability, mechanism, or developmental constraint, which are progressively given operational definitions and find a new and more important theoretical place at the basis of new empirical research programmes in neurobiology, molecular biology and evo-devo, respectively. In the following section, we focus on four concepts the meaning and potential fecundity of which are discussed in the contributed papers, namely the concepts of organism, interactionism, evolutionary novelty (including related concepts like phenotypic plasticity, directed mutations, or pre-adaptation), and adaptation. We suggest that stripping the last concept of its controversial features would be profitable for future biology, and that clarifying the meaning of the first three ones will probably allow for the building up of new research programmes.

1.2 Elaborating Key Concepts

Throughout the contributions of this volume, four key concepts emerge as potentially playing a major explanatory role in future biology, a role even greater than the one they play currently: the concepts of organism, interactionism, novelty in evolution, and adaptation. Not all these concepts are new; their meanings and implications are evolving and they make up together a new image of life sciences.

1.2.1 Organism

A striking example of the fecundity of the ongoing reflection on biological concepts is the renewed interest in the role played by the concept of an organism in developmental biology as well as in evolutionary biology. According to Lewontin (1983b), and contrary to Dawkins (1976), the concept of an organism should be put at the centre of evolutionary theory. Dawkins initially claimed that organisms are too transient to be of any evolutionary importance, whereas genes, conceived as everlasting replicators, are the real objects of evolutionary theory. Nevertheless, working out the "gene's eye view" led Dawkins to a later "rediscovery of the organism" in the guise of an "extended phenotype" (Dawkins 1982; see also the section called "The Organism Strikes Back" in Sterelny and Griffiths 1999). Dawkins' concept of an extended phenotype does not resort to the usual phenomenal properties of organisms but is rather defined in relation to the potential evolutionary effects of an organism's genes. Turner (2000) further emphasizes how important the effects of organisms' agency are from an evolutionary point of view.

Lewontin (1983a,b) strongly insisted on the necessity of conceiving organisms as co-evolving with their environments, and as acting on them, rather than as passively submitted to external environmental selection pressures. Inspired by many aspects of Lewontin's thought (but see Sterelny 2001), proponents of Developmental Systems Theory have built up a grand conceptual framework aiming at apprehending organisms in all of their aspects: as developing from birth to death and as agents, engineering their environment and co-evolving with it. They even see the organism as only a part of more general biological units, namely "developmental systems" (Oyama) or "life cycles" (Griffiths and Gray).

In developmental biology, the core character of organisms is clear: the organism seems to be the focus of every explanatory or modelling project related to development. The organism is best viewed, from this perspective, as "the object that allows us to integrate many different dynamical processes, such as development, genotype-environment interactions, epigenetic effects and inheritance patterns, behaviour, constraints, etc." (Laublichler, talk at Paris 2006 workshop).

What all these conceptions have in common, despite significant differences, is the notion that the concept of an organism should be defined via its activity, namely via biological interactions, and not via its visible borders, like membranes, skin, etc. As Wimsatt claims (2001), the phenotype and the organism's activity have to enter population genetics models in one way or another. Usually, they enter models via implicit and unanalyzed assumptions. However, the models will probably be improved when these assumptions are conscious and controlled. Wimsatt claims that DST is the right theoretical framework to do that. We think that the texts gathered here demonstrate that attempts to put into question the role of the organism in evolution, development, and other key biological aspects have now proven misguided.

Our conviction is that in the near future the organism will be the centre of a major reconceptualization, based on the notions of agency, co-construction with the environment (including niche construction), and phenotypic plasticity. Lewontin's hopes about the recognition of the evolutionary and physiological significance of organisms seem close to be fulfilled.

1.2.2 Interactionism

According to Richard Lewontin, the interactions between organisms and their environments are still currently underestimated in evolutionary studies. However, the very nature of these interactions should not be taken for granted but has to be carefully scrutinized beforehand. Two extreme views should indeed be avoided, labelled "internalism" and "externalism" by Lewontin (for an elaboration on these terms, see Godfrey-Smith 1996). According to internalism, the organism is the main agent of its relationships with its environment, whereas according to externalism, environment holds a shaping power on organisms' traits and behaviours.

In order to take Lewontin's point into account, several proposals have been made toward more "interactionist" variants of the theory of evolution (Lewontin 2000b). The term "interactionism" can take a variety of meanings, carefully analyzed in Oyama (2001). As Oyama clearly shows, just resorting to weakly defined "interactions" between organisms and their environments is not enough to do justice to Lewontin's complaint. Consequently, several other concepts have been suggested: Susan Oyama favours *constructive interactionism*, the recent studies on niche construction are often labelled *co-constructionist*, while Lewontin himself prefers referring to the *dialectical* relation between the organism and its environment (Lewontin 1983b, 2000a,b). Only these "constructive", "co-constructionist" or "dialectical" variants of interactionism fully account for what is lacking in today's evolutionary studies according to Lewontin. Still, such a major gain in conceptual clarity is only a first step toward the elaboration of empirical research programmes. How to develop such research programmes is a major theme of this book (see in particular the contributions by Odling-Smee and Gilbert).

1.2.3 Novelty in Evolution

The contributions of Gilbert, Odling-Smee, and Sterelny in this book investigate various ways of articulating developmental and evolutionary biology. They suggest differentiated answers to the question: How far has evo-devo come today? They all agree that the needed articulation of developmental and evolutionary biology, namely the hope for a deepening of the evo-devo synthesis, will likely focus on the question of the origin of phenotypic novelty. "Novelty comes from random mutations" is the traditional answer to that question, coming from the view that all evolutionary changes originate in modifications in genes. This genocentric view has been vigorously challenged by Oyama (2000b [1985]) Oyama et al. (2001), West-Eberhard (2003), Jablonka and Lamb (1995; 2005), Gilbert (2006a)... to name a few. As Gilbert et al. (1996, p. 361) have it, "Microevolution [i.e. evolution studied from the traditional, population geneticist's point of view] looks at adaptations that concern only the survival of the fittest, not the arrival of the fittest" (see also Gilbert 2006a). According to the challengers to the genocentric view, novelty can come from many other places than the genes. Phenotypic plasticity and the modularity of the developmental processes are seen as major providers of novelty. Because of its capacity to induce evolutionary change, environment itself can be a genuine source of novelty as well. As a result, a brand new "eco-evo-devo" synthesis is arising, trying to articulate phenotypic modifications induced by environment (sometimes "engineered" by organisms themselves via niche construction) and their evolutionary effects (Gilbert 2001). Whereas the relative importance of niche construction in evolution has been recently discussed (Odling-Smee et al. 2003; Sterelny 2005; Laland et al. 2005), studies on phenotypic plasticity (West-Eberhard 2003; Sterelny, this volume) are on the edge of being integrated into wider reflections about the actions of organisms on their environments (Odling-Smee, this volume).

1.2.4 Adaptation

In a provocative paper, Gilbert et al. (1996) have claimed that the role of natural selection is "merely a filter for unsuccessful morphologies generated by development". They want to convince their readers that "a tertium quid, development, has been imposed between Mayr's two categories of functional and evolutionary biology". Michael Ruse has vehemently attacked this view by claiming that natural selection is *the* key to adaptation and novelty (Ruse, this volume, answer by Gilbert (2006a), answer to this answer in Ruse (2006c)).

Surely, our views about adaptation have been enriched by numerous debates during the last decades, a milestone of which being Gould and Lewontin's 1979 paper (see also Godfrey-Smith 2001a). Nevertheless, an almost hidden problem facing this notion, emerging from the papers in the last two parts of this volume, is the question of what exactly is adaptive to what. Organisms to their environments? Phenotypic traits? Developmental mechanisms? Life cycles? This question is seldom asked, generating much confusion. A fruitful framework to raise it has been provided by Depew and Weber (1995): according to them, organisms have first to be conceived of as autocatalytic dissipative structures, which implies that every question concerning adaptation has to be answered by resorting to organisms' autocatalytic capacities. If one conceives of organisms this way, natural selection is "presumptively adaptive" (Depew and Weber 2001): selected traits cannot be but adaptive. This is one of the radical options in the debate. Michael Ruse, in this volume, is close to this option, for albeit different reasons. Scott Gilbert, on the other hand, defends a less radical conception of the relationships between natural selection and adaptation, more sensitive to the phenomena that may delay or modify its action.

1.3 Detailed Contents of this Book

The first three chapters are an enquiry into the epistemological obstacles which today have to be overcome in biology. They show why the articulation between molecular and evolutionary explanations is a difficult process. The process of abstraction that each explanatory scheme generates – with the production of models and new entities – frequently leads to a reification of these entities, i.e. their conversion into "objects" which no longer need to be defined. The existence of such "hardened notions", specific to each explanatory scheme, makes this articulation even more difficult. A concept like that of information, which has been so influential in the rise of molecular biology, is a source of numerous ambiguities and opens the door to metaphysical conjectures.

Michel Morange (Chapter 2) first displays a sharp contrast between rapidly developing domains, like genome sequencing programs, the structural determination of macromolecules, and the elaboration of sophisticated mathematical models on the one hand, and the fragility of the underlying conceptual framework of today's biology on the other hand. His main thesis is that we are suffering of the epistemological and sociological difficulties to link molecular explanations with those generated by other lines of research, like evolutionary, epigenetic, or self-organizational (physical) explanations. He points out some cases is which these difficulties have been overcome as leading examples for the future of biology.

Susan Oyama (Chapter 3) remarks that, among the biologists and philosophers who have found the notion of genetic information indispensable, several are famously and vocally antireligious, but at the same time share their devotion to infotalk with certain scientifically inclined theologians writing about evolution. According to her, this, in fact, should not surprise us, for minded matter is just what we get when information is construed semantically. The religious evolutionists share with their atheist adversaries a commitment to what amounts to molecular intentionality. She argues that this shared commitment is neither coincidental nor benign, and raises the question of why information has been so attractive to so many.

Peter Godfrey-Smith (Chapter 4) takes on one of the most fundamental epistemological issues underlying the questions dealt with in this volume, namely the nature of the operations of abstraction and idealization applied to evolutionary biology. He first defines these two operations and then examines a typical error coming from an operation of abstraction: the reification of its results. He shows that talking of genes as de-materialized, informational entities originates in such an error and argues that this move is unnecessary and misleading.

Chapters 5, 6, 7, and 8 focus on one of the domains where the interweaving of approaches and conjunction of technical tools has been recently the most successful: the wedding of evolutionary synthesis and developmental biology. We will consider it in an even broader meaning, Eco-Evo-Devo, including the studies on the complex relations between environment, evolution and development. The simple relation and clear separation between genotype and phenotype has been clearly challenged. There is a huge plasticity of the phenotype, resulting from the environmental influences on development. The complex interactions of the organism with the environment modify the latter, creating new selective conditions for the genotype. This transformation of the relations between genotype, phenotype and the environment leads to the adoption of new animal models by biologists, better adapted to the new vision. At the same time, the proliferation of studies in developmental biology leads to improved descriptions, but also to an accumulation of data, the use and interpretation of which is still problematic. In general, this field of research is characterized by a huge richness in interactions and conceptual novelties, and constitutes a laboratory for the future of biological sciences, a privileged place for philosophers to scrutinize the evolution of concepts and models.

In his paper (Chapter 5), Scott Gilbert emphasizes that whereas evo-devo has been characterized by the revolt against model systems, today's research needs new model systems for evo-devo and for eco-devo. He gives some examples of these new model systems and points out what is at stake in the choice thereof, namely understanding what constitutes normal development and how macroevolution occurs. Avoiding biases is a very important issue, especially when it is realized that the environment plays an instructive role in phenotype production, so that the developmental program is dispersed and that model systems have to be seen as genotypes + spatio-temporal context of organisms. Niche construction thus appears as a decisive component of future eco-evo-devo.

John Odling-Smee (Chapter 6) suggests that his work on niche construction might lead to a fuller synthesis of evolutionary biology, ecosystem level ecology, and developmental biology than the incomplete one that is available. He insists on explicitly recognising niche construction as an approximately equally potent and co-directing cause in evolution working in conjunction with natural selection, generating feedback in evolution caused by the modification of environmental natural selection pressures by the actions of organisms. He clarifies the way in which niche construction can provide a powerful "evo-devo" link by exploring the consequences of ecological inheritance and "internal niche construction" on developmental processes.

In his study of "the influence of phenotypic variation on evolution" (Chapter 7), Kim Sterelny first insists that the set of all possible variations in a given lineage is not homogeneous: as a structured set, it influences the evolutionary trajectories of the lineage. He thus opposes a classic conception of Darwinism in which the supply of variation is isotropic. Sterelny then emphasizes the central, pre-adaptive role of phenotypic plasticity in understanding evolvability. Plasticity may be coupled with environmental induction and genetic accommodation to generate evolutionary novelty. Finally, Sterelny shows how niche construction, in the broad sense of niche choice as well as in the narrower sense of niche engineering, is linked to the evolution of novelty via the "organism-environment developmental loop". In the final part of the paper, he explores the hypothesis that the organisms in a given lineage can influence the supply of variation to selection and hence determine the lineage's evolutionary trajectory.

Chapters 8, 9, and 10 are devoted to the treatment of this complexity underlined in the previous part. Complexity is a recognized characteristic of organisms. Nevertheless, adaptation by natural selection and theories of self-organizing complex systems are often seen as opposed explanations of the observed phenomena. The origin of life is clearly one of the domains where the confrontation between the different interpretations is the most obvious. But in contrast to the Evo-Devo domain, this field appears more heterogeneous, the oppositions acute, and the equivalent of the Evo-Devo synthesis still awaiting to be born.

In his paper (Chapter 8), Mark Bedau starts from the observation that the evolution of the biosphere exhibits a trend of increasing maximal organism adaptive complexity. He then insists that the "arrow of complexity hypothesis", namely the hypothesis that this dynamical emergent process is a generic property of some class of evolving systems, rather than just an accident, has to be recognized as a genuine empirical hypothesis, open to creative investigations. His main claim is that the constructive, emergent models of soft artificial life are the adequate objects to investigate in order to assess the truth of the arrow of complexity hypothesis. Although these models (e.g. Tierra or Avida) do not exhibit by now any relevant increasing complexity, Bedau claims that it is by constructing new models, endowed with yet-to-be-discovered evolution mechanisms, that we shall solve this question.

Evelyn Fox Keller (Chapter 9) explores the meaning of self-organization when it applies to the emergence of living entities. She argues for an expanded conception of self-organization including self-assembly or composition and claims that it allows for a better understanding of life before the advent of natural selection, namely life submitted to the "survival of the most robust" principle. According to her, evolution by composition (pace strict gradualism in pre-Darwinian evolution), which may be fortuitous, may explain the emergence of cellularity and of function. She therefore argues against the idea of a continuous increase of either complexity or robustness, insisting on the abrupt transitions occurring before natural selection. In the second part of the paper, she tries to dismiss the "gospel of inevitability" by arguing, contrary to Kauffman, Depew and Weber, that the emergence of life may be given a physical explanation even though we do not see it as highly probable.

In Chapter 10, Michael Ruse defends a strong neo-Darwinian credo. He contrasts the claims of two self-proclaimed challenges to contemporary evolutionary theory with a well-understood neo-Darwinism. His two targets are first the idea that developmental constraints limit the role of natural selection (attributed by Ruse to Gould and Lewontin, and to Gilbert, Opitz and Raff), and second the thesis that, at least in some cases, evolution is due to self-organization and not to natural selection (a thesis attributed to Goodwin and to Kauffman). Against these two views, Ruse asserts that:

(i) the "organized complexity" (adaptedness) of the living world has indeed to be taken seriously: it is what natural selection aims to explain.
(ii) self-organization has nothing to say about this organized complexity: there is no such thing as "order for free" in evolution, he believes, since natural selection does all the job.

Acknowledgments We are deeply grateful to Susan Oyama, Hannah-Louise Clark and Marie-Claude Lorne† for their corrections and comments on previous drafts.

Chapter 2
Articulating Different Modes of Explanation: The Present Boundary in Biological Research

Michel Morange

2.1 Introduction

For an observer totally unfamiliar with the biological sciences, the present picture of these disciplines would be highly contrasted. On the one side, he would see an accumulation of new data provided by genome sequencing programs and the structural determination of macromolecules. Whereas the interpretation of the former is difficult – but opens to a very fruitful comparison between organisms -, the second leads in most cases to unambiguous explanations of biological phenomena. On the other hand, he would also observe in ecology and evolutionary biology the proliferation of sophisticated models that require highly complex mathematical tools and either raise "big" biological issues, such as the meaning of sex or the significance of the ageing process, or conversely focus on the explanation of "local" facts, the characteristics of such and such an animal and plant population, and their recent evolution.

At the same time, if such an observer is receptive to the discourse of philosophers, he will be surprised by the fragility of the conceptual framework which guides this plethora of work. Basic concepts such as gene (Pearson 2006) or program do not have any simple content or meaning. The general use by biologists of the informational concepts is looked at with scrutiny. The requirement for an informational language is debated, and the risks and ambiguities to which it leads are regularly outlined. Traditional approaches in biology are challenged by new ones: genetics by epigenetics and developmental systems (DS) theory; the molecular description of the genetic program by self-organization models; the explanatory power of natural selection by the capacity of biological systems to self-organize; and the Darwinian model by apparent Lamarckian revivals.

In parallel, new models are emerging from bioinformatics and mathematics, supported by an increasing number of mathematicians, physicists and bioinformaticists who are entering the field of biology. A lot of new words are introduced to designate emerging fields. These new disciplines have an open future, but some of them, such

M. Morange (✉)
Professor of Biology, Ecole normale supérieure (ENS) and Paris 6 University, Paris, France
e-mail: morange@wotan.ens.fr

A. Barberousse et al. (eds.), *Mapping the Future of Biology*, Boston Studies in the Philosophy of Science 266, DOI 10.1007/978-1-4020-9636-5_2,

as synthetic biology or systems biology, are already the object of active discussions between biologists and philosophers (O'Malley and Dupré 2005).

Probably the best way to organize this chaotic landscape – and the one for which I am the best equipped – is to consider its evolution as a response to the tremendous transformation of biology that has occurred since the 1960s and 1970s with the rise of molecular biology. But instead of rejecting the explanations of this new approach, as most commentators do, I will show that the main current trends in biology are the results of efforts to link molecular explanations with those generated by other lines of research. Finally, I will discuss the difficulty of linking different explanations, its limits, and what it may tell us about the future of biology.

2.2 Molecular Explanatory Models

The heterogeneous array of models and theories aimed at explaining the biological facts can be ordered if one admits that most of them were developed to overcome the limits and weaknesses of molecular explanations. Molecular explanations have evolved over the last forty years: the molecular models of the 1970s were more oriented towards DNA and genes than the models of today. I will briefly describe the molecular explanatory framework of the early 21st century, knowing perfectly well that it is a highly schematic presentation.

Molecular biology is not only a very rich array of techniques, but also a physico-chemical explanation of biological phenomena. Biological processes are considered to rely on the molecular functions associated with individual macromolecules, proteins in most cases, but also RNAs as recent studies have clearly outlined. Explanations of the molecular functions can be found in the very precise description of the structure of the macromolecules responsible for them. Explanations of biological processes emerge from the conjunction of individual molecular functions. DNA – the genetic material – contains the information to synthesize these macromolecules: not only the information necessary to add one nucleotide or one amino acid to another, but also part of the information required to regulate gene expression, i.e. the level at which the macromolecules are synthesized.

Some processes discovered during the last two decades blur this landscape: the information contained in the genome can be modified, by editing or splicing, generating from one simple segment of the genome a family of proteins, instead of a unique one, having different properties, or a protein having a slightly different structure from that anticipated. I consider that we can leave these observations aside. Their importance remains limited, and they do not profoundly alter the explanatory models of molecular biology.

Consider now how the explanatory model of molecular biology was challenged. The first criticism came from evolutionists, and in particular from Ernst Mayr. In 1961, in an article entitled "Cause and effect in biology", he argued that proximal explanations, as those provided by molecular biologists, were not sufficient, and that they had to be complemented by the teleonomic explanations of evolutionary biology (Mayr 1961). Taking an example from his early scientific career, Mayr argued that the explanation of the migratory behavior of birds could not be limited to the physico-chemical description of the mechanisms by which the animals started to migrate. The purpose of the behavior, the increase in fitness that it gives to those who adopt it in specific environmental conditions, which have to be described, is a major feature of the overall explanation. For Ernst Mayr, only teleonomic explanations are specific to biology, since molecular explanations are physico-chemical and therefore not limited to the biological sciences. Mayr was at least partially right: evolutionary explanations have resisted the rise of molecular biology quite well.

Most criticisms of the explanatory value of molecular biology targeted the genetic and informational aspects. The first set of criticisms concerns the possibility of deducing the complexity of biological processes from the information contained in the genome. The second set of criticisms targets the possibility of limiting the transfer of information from one generation to another to the transmission of DNA. Let us consider these two limits to the explanatory power of the molecular models separately, although the arguments are frequently intertwined.

Efforts have been made to demonstrate that the information contained in the genome is insufficient to generate the complex characteristics of organisms. Due to the limits of a quantitative estimate of information, these attempts were not generally recognized as significant. Nevertheless, the idea that there was something more than the simple interaction and self-assembly of macromolecules was shared by most critics of molecular biology. But the nature of this “more”, its epistemic or ontological nature, is the subject of active and confused debate.

The first idea is that there are some self-organization principles that guide the assembly of these molecular components. The stochastic variations that affect these molecular components are frequently considered as the origin of these self-organizing processes. The motor of the self-organization process may also lie, not in the system itself, but in the relation between the living system and its environment. The thermodynamic imbalance between the system, the organism, and the environment can generate order through the formation of dissipative structures. Order can simply emerge from the non-linearity of equations describing the system and the relations between its components, which generates characteristic dynamic behaviors.

It is clear that the nature of the limits of the molecular explanations will vary from one point of views to another. Epistemic in the latter model, it has an ontological flavor in the first two models.

My presentation is obviously a little cursory. The various models of self-organization and complexity deserve to be studied more precisely, and they are in the other contributions.

A second criticism of molecular explanations is the impossibility of reducing hereditary transmission to genes and their informational content. Heredity is more than the genetic code. Genetics is only one part of a general process of transmission through the generations.

The limits of the genetic paradigm of hereditary transmission are multiple. On the one side, it is easy to demonstrate that, during reproduction, structures other than DNA are transmitted. Reproduction of organisms from one generation to the next is not the simple transmission of a DNA molecule, but a complex process of fusion/reproduction of cells: membranes, organelles as mitochondria, are also transmitted down the generations.

On the other side, hereditary transmission in the human species is not limited to DNA and other molecules. Behaviors, and even a certain form of environment, are also inherited. Cultural and environmental heredity is not restricted to human beings and other mammals, but found everywhere in the living world, from insects to birds. Such examples have been fully investigated in DS theory (Oyama et al. 2001).

The role of epigenetics in this general landscape is multifaceted (Jablonka and Lamb 2005). Epigenetic mechanisms are responsible for the transmission of organismal characteristics through the generations, independently of the transmission of a specific DNA sequence. DNA methylation and chromatin structure can strongly affect gene expression and be transmitted, at least partially and transiently, and in certain organisms, from generation to generation. But epigenetics also refers to the fact that the construction of the organisms is more than the simple addition of its components – at least in the first meaning given to the word by Conrad Waddington (Waddington 1942). Epigenetic studies frequently underline the limits of genetic information – for instance in the precise design of neural pathways – and the role of the complex interplay between the organism and its environment.

I have tried to present the different conflicting models by showing that they aim to highlight the limits and weaknesses of molecular explanations. This explains part of the heterogeneity of the field. For instance, epigenetics aims to show the limits of genes both at the level of the construction of organisms and in the hereditary transmission of characteristics; whence its multiple – and confusing – meanings.

Some of the discussions might be made clearer by precisely defining what is at stake. For instance, epigenetic mechanisms can explain how the level of expression of genes can be transmitted through generations; they are unable to produce the transformation of one sequence into another. Whatever its future extension, epigenetics will never be able to replace genetics.

As we have seen, one major problem is that molecular explanations and self-organization models do not operate at the same level: the former aims to characterize very precisely what happens in cells and organisms; the latter to provide general guiding principles for the organization of the system. A further difficulty is that the dichotomy introduced by Ernst Mayr between molecular and evolutionary explanations is rejected by those proposing self-organization models. For the latter, this dichotomy has no "raison-d'être", the principle of self-organization being at the root of both the development of organisms and their evolution.

2.3 Putting These Models Side by Side or Opposing Them is a Dead-End

Putting these models side by side and simply considering that different explanations have to coexist in biology is not an appropriate solution. For in this case one has to consider that the models are "orthogonal" one to the other, i.e. independent. This was the point of view adopted by Ernst Mayr. In a text written for his 100th birthday, Mayr wrote that tremendous transformations affected biology from the 1930s with the rise of molecular biology and genomics, and yet the evolutionary synthesis remained unaltered (Mayr 2004). This is obviously untrue. The importance of neutralism (Kimura 1968), and the recent development of evolutionary developmental biology, or Evo-Devo, are the indirect consequences of the development of molecular biology. The same is true for the so-called revival of Lamarckism in biology: the evidence for the capacity of organisms to alter their rate of mutation – of the whole genome or of parts of it – in response to variations in the environment and, in particular, in stress conditions. These mechanisms are not Lamarckian – the mutations are random and not adapted to environmental variations – but they obviously extend the Darwinian evolutionary model. They establish a quantitative link between variation and selection, the former being increased in conditions where the second is expected to be harsher: the situation is conceptually similar to what happens with the notion of niche construction; the distinction between variation and selection is not abolished, but the two processes are related by a novel link. The evidence for such mechanisms of control of gene variation rate and their characterization are the result of molecular studies on the mechanisms by which cells repair the DNA molecule.

And in a symmetrical way, molecular biology was deeply marked by Darwinism. Consider, for instance, the central dogma of molecular biology – that it is impossible for information to transit from proteins to nucleic acids – which finds its full justification not in experiments but as a direct translation into molecular terms of neo-Darwinian convictions.

The passive coexistence of models is not possible, but opposing them is no better. The development of more and more powerful technologies allowing fast, high-resolution determination of macromolecular structures, and the evidence for the great explanatory power of this structural information, is completely at odds with the pretension of the advocates of some of these alternative approaches to replace the molecular description altogether. In addition, these new approaches are often narrowly linked with the previous molecular vision. Consider, for instance, epigenetics. Setting aside the fuzziness of this word, and considering the work done in this field today, it is obvious that epigenetic phenomena – methylation, organization of chromatin structure – cannot be described and understood independently of genetics. The prefix "epi" of epigenetics must be taken seriously. Epigenetics is above genetics: epigenetic phenomena can only be described and find their place through the alterations and modifications they bring to genetic phenomena. Epigenetics cannot exist without genetics. And the lesson is probably of general value: the future of biology is neither in the passive coexistence of different explanations nor in the replacement of one by another, but in the precise interweaving of these different explanatory schemes.

In a similar way, long overdue is the linking of the molecular and evolutionary explanations. The coexistence of molecular descriptions explaining the function of molecular modules with no reference to their evolutionary history, and of evolutionary models paying no attention to the nature of the components of the system, is no longer possible. Natural selection is blind to the precise nature of the systems on which it acts (Rosenberg 1994), but this does not mean that it is indifferent.

Precise knowledge of developmental genes gives clues to what routes are open to evolution (Kirschner and Gerhart 2005). In the very active field of research of Evo-Devo, researchers try to relate the action of developmental genes and their alterations to evolutionary transformations. Instead of considering as in Modern Synthesis that all genetic variations are somehow equivalent in their amplitude and therefore on evolution, the specialists of Evo-Devo focus their studies on the variations of developmental genes, which may have dramatic effects on the development of organisms, and on its functioning and morphology in the adult stage. The molecular characteristics of developmental genes do not guide evolution, but they do impose a specific topology on the possible evolutionary landscapes. The case of the Cambrian explosion, the apparently explosive emergence of new animals five hundred million years ago, is exemplary. The early efforts of molecular biologists were aimed at explaining the Cambrian explosion by the "invention" of developmental genes. Molecular dating showed unambiguously that the appearance of developmental genes – and, in particular, homeotic genes – largely preceded the Cambrian revolution. The present hypothesis is that its origin has to be looked for in some not yet well-defined variations of the environment which allowed these diverse animal species to occupy new niches. But the richness of the body plans thus revealed would have not existed without these developmental genes.

This example illustrates also what might be the place of Evo-Devo. It must not have the ambition to replace the Darwinian explanations. The trigger to evolution has to be looked for in the adaptation of organisms to their environment, not in the invention of new – developmental – genes. But once a new niche has been created, the characteristics of the organisms which will occupy it are the result of the action of these developmental genes.

The dovetailing of molecular and evolutionary explanations will give a precise meaning to the vague notion of constraints. It will lead molecular biologists to pose new questions in the description of the systems they study. Not only to explain why these systems work, but also why they have been selected in preference to other systems, and why they have such and such characteristics. The explanations will not be universal – the characteristics of a system can be the fruit of an optimal adaptation, or the result of a frozen accident – but they are worth looking for.

And this is precisely the direction taken by recent work on systems that have been already extensively studied by molecular biologists. The best example is probably the case of the lactose operon. This is the genetic regulatory system which has been the most extensively studied, and the best characterized. But simple evolutionary questions concerning this system were never raised. Is the system optimally adapted to the concentrations of nutrients that the bacteria encounter during the course of their lives? Can we demonstrate that? (Dekel and Alon 2005). Other solutions were probably capable of enabling the micro-organisms to adapt to these varying concentrations of nutrients. In particular, simpler mechanisms of stochastic variation of gene activity are used by pathogenic organisms to adapt to the new environment represented by their hosts. Why was the complex, expensive responsive lactose system preferred to these simpler mechanisms? (Kussell and Leibler 2005) What kind of benefit does it provide? All these questions were raised in recent publications and have received preliminary answers. They constitute a form of interplay between molecular and evolutionary explanations which did not exist before.

The interlinking of molecular and evolutionary explanations is probably the best way to limit the space afforded the sometimes illusory explanations provided by the models of self-organization. It is obvious that these models have frequently flourished in the vacuum that has existed between the molecular and evolutionary explanations. But the models of self-organization have also found a favorable ground in the rift between the molecular and physical explanations. To say that organisms are physical systems which obey the laws of physics is a statement of the obvious. But to interweave molecular and physical explanations is a much more difficult enterprise which has only recently been initiated.

Maybe the most significant example is to be found in the very active field of research on molecular noise. Molecular noise is the expression used to designate the stochastic variations in molecular processes, in particular in gene expression, generated by the low number of molecules involved and the slow rate of some of the reactions taking place inside cells. The study of molecular noise has been neglected by molecular biologists for different reasons. The first – good – reason was the lack of techniques to study it. This technological obstacle was overcome by the development of molecular tools to fuse the proteins under study with reporter proteins and the use of very sensitive physical devices to detect the signals emitted by these reporter proteins. The "bad" reason was a lack of interest in molecular noise, through ignorance of its origin and nature, and because of an undue admiration for the precision and reliability of molecular processes operating in organisms.

The number of studies devoted to molecular noise has exploded over the last years (Raser and O'Shea 2005). Most of these have simply been aimed at describing the nature of molecular noise, its amplitude, and its origin. Some work is though already focused on describing how the organisms manage molecular noise, how the architecture of the networks limits its consequences, and conversely how the organisms can exploit it to generate transient diversity likely to improve their adaptation to varying conditions, or to give rise to complex structures during development. An explanation may be at hand for the origin of phenotypic plasticity, which is so important in evolution. Such studies dissipate the clouds under cover of which some models of self-organization have given noise a pre-eminent role. The truth is simpler: molecular noise exists, and organisms have learnt to deal with it, and possibly exploit it.

A second example, once again focused on the lactose operon, underscores the value of merging different explanations to generate a full description of a biological system. Like many other regulatory systems, the lactose system has a positive feedback loop – the membrane protein allowing the import of lactose into the cell is itself induced by an increase in the intracellular concentration of lactose – and a bistable mode of functioning. The existence of bistable – or multistable – states of functioning is reminiscent of the phase transitions extensively studied by physical chemists. It is a characteristic of many biological systems, and always linked with the existence of such positive loops. But the reverse is not true for reasons which remained obscure until the system was modeled. The model revealed that some parameters are crucial for the existence – or otherwise – of bistable behaviors. The predictions of the model were tested and confirmed by using targeted mutagenesis to alter the characteristics of the molecular components of the system (Ozbudak et al. 2004).

In contrast, it is probably not a coincidence that the models of self-organization today take center stage in the scenarios of the origin of life. The scarcity of "hard facts", the difficulty of linking together molecular physico-chemical explanations and evolutionary models, in the context of the prebiotic soup – whatever its precise nature is – leave considerable room for models of self-organization.

2.4 The Difficulty of Interlinking Different Explanations

The huge efforts required to link explanations rooted in different fields of research are frequently explained by the well-known difficulties of inter-disciplinarity. This explanation has two limits. First, it omits to mention the difficulties of joining up explanations within one discipline – the case of molecular and evolutionary explanations is one of the best examples, but others can be found in other disciplines. Second, it considers that the origin of the difficulties is mainly sociological, whereas its roots are both epistemological and sociological.

The major obstacle to such merging of explanatory schemes is that with the increasing specialization of scientific careers each scientist has been trained to favor one explanatory scheme. Any increase in efficiency in using this explanatory scheme is paralleled by a decreased aptitude to use other explanatory schemes, and even to appreciate their interest. This leads to the artificial opposition of explanatory schemes which obviously are of a different nature, and therefore fully complementary. One striking example is the different explanations given for the influenza pandemic which in 1918–1919 killed more European and American citizens than the battles of the First World War. Traditionally, it has been considered that the influenza spread because people were debilitated after four years of war, and because the movements of populations and troops favored the appearance of an epidemic, and its transformation into a pandemic after the war. The recent molecular characterization of the influenza virus responsible for this pandemic has pointed to another explanation: the virus had specific molecular characteristics which favored its passage from its traditional host – avian species – to the human being, and its aggressive development in the latter (Tumpey et al. 2005). This example is emblematic for two reasons. The first is that the two explanations obviously should not be opposed or compared, but rather linked. The second is that each advance in one particular scientific description – in this case the molecular description of the virus – immediately tends to favor the explanatory schemes associated with this description.

Another obstacle to the merging of explanations stems from the "spontaneous epistemology" of scientists. Exceptions aside, scientists are generally not interested in philosophy of science or philosophical disputes. Nevertheless, some 20th century debates in philosophy of science have permeated the scientific community. That scientific theories can be falsified, but not verified, or that science consists of a succession of paradigms, is "part and parcel" of the tacit knowledge of scientists. But scientists tend to remain in ignorance of the criticisms addressed to the falsificationnist model of Karl Popper – with a distinction drawn between a central core of knowledge and a protective belt -, or of the doubts expressed regarding the precise meaning of the term "paradigm", and of the difficulty of extending this notion beyond physics. The simple lesson that has been learned is that the main objective for scientists should be to falsify a well established theory or to introduce a new paradigm. Scientists' abuse of the notion of paradigm has already been underlined. The degenerate form of epistemological knowledge that has permeated science is used as an argument for intolerance and rejection.

In a similar way, some epistemological debates absorb all the attention of scientists while masking more interesting issues. Such is the case of the reductionist/holistic debate on the future of molecular biology, to which one should add the place and significance of "emergent" phenomena. Our intent is not to deny the intellectual interest of these debates, but serious bias does arise if this question is the sole focus of discussion of what happens today in the field of biological research. To reduce the present state of biology to a transition from a reductionist to a holistic vision of biological phenomena does not acknowledge the richness of the studies being presently done. It prevents us from seeing that what is at stake is a search for a way to link different explanatory schemes. The most active works pursued today in systems biology does not seek to replace the molecular description by a holistic one, but rather to link a molecular description to another one – in terms of the structural and dynamic properties of networks – located at a different level of organization (Guido et al. 2006).

Efficient linking can only be achieved through precise experimental systems and objects, and not by general considerations. And the absolute condition for a dovetailing of explanations is an equality between partners: the system has to be of an equal interest for both. There are a limited number of experimental situations in which the system is of equal value for scientists of different disciplines who use different explanatory schemes. The existence of an inequality is probably the major source of failure in interdisciplinary research: one discipline imposes its own experimental system – generating a question which will be considered by the other as peripheral to its own interests, and of a limited value. Much of the work on self-organization has been focused on the generation of forms in animals, whereas most molecular biologists consider forms to be just a secondary consequence of internal molecular organization. The use of mathematical and/or physical models to explain the layout of stripes on animal bodies, or the complex pattern of colors in plants is, at best, considered anecdotal by molecular biologists.

In addition to being of interest for the two potential collaborators, a system has to be accessible to the technologies and tools used by both. Once again, such optimal cases are rare. Most of the models of population geneticists focus on characteristics – behaviors, life histories – the molecular basis of which is totally ignored. The most interesting regulatory networks, such as those involved in the control of cell division and in its dysregulation in cancer, are by their very complexity beyond the reach of a physical and modeling approach.

2.5 A Paradoxical Conclusion

Some of the most interesting recent studies can be described as attempts to link research efforts pursued within three hitherto quite separate explanatory schemes: molecular, evolutionary, and physical. I previously described the studies on the lactose operon which are at the boundary between the molecular and evolutionary approaches. Modeling of the system, and explaining the origin of the bistable mode of functioning, are at the boundary between the molecular and physical approaches. Trying to find the origin of the robustness of networks in their topological characteristics is a way of interlinking the physical and Darwinian traditions.

The present study clearly has its limits. I could add to my list of attempts to link explanatory schemes many others, for instance a recent study that links the stabilization of sexual reproduction with the properties of gene networks (Azevedo et al. 2006). Existing explanatory schemes need to be characterized more precisely, and other schemes may well emerge and complicate still further attempts to draw them together. Yet despite these limitations, my take-home message is unambiguous. The future of biology does not lie in the development of a new explanatory scheme, or in the disappearance of some schemes in favor of others, but in an active interlinking of different schemes. The rules governing this interlinking will have to be fixed for each study. Does such a model help us to understand the present situation? Does it tell us something significant about the future of biological research?

My feeling is that the apparent consistency of the self-organization and complexity models was only due to the huge distance which has hitherto separated the explanatory schemes that coexist in biology. As explanatory schemes are linked, these models will appear more and more heterogeneous and fuzzy, and vanish. A more positive view would be that they will progressively dissolve – or be incorporated – into the numerous studies that will be done at the boundary.

In a similar way, I see the importance of DS theory and of studies on epigenetic phenomena more as the result of the rift between the explanatory schemes than as a sign of the emergence of a new, global approach to biological phenomena.

This does not mean that the emphasis placed on epigenetics, self-organization models and DS theory has been useless in the recent history of biology. It was a way to point to the vacuum that remained between the different explanatory schemes, and their limits as long as they remained isolated, each having claimed to be the sole and sufficient explanation of biological phenomena. But the epigenetic and self-organization models were too heterogeneous to constitute a new approach to biological phenomena. As some supporters of DS theory admit, their models were at the border between science and philosophy, and probably more on the side of the latter: they were not, in most cases, research programs, alternatives to the present models and theories awaiting for a paradigm shift. Epigenetics is already a recognized field of research, enriching the genetic description; self-organization models will be used to enlighten such and such aspects of the complex biological systems; and the numerous original prospects opened by DS theory will be progressively incorporated into an extended form of Darwinism. We can illustrate the future contributions of these models and theories by using one simple example from the past. The understanding of the functions of the nervous cells, and in particular of the conduction of the nerve influx by the sophisticated and highly controlled movement of ions through channels across the cell membranes, has been illuminated by the use of concepts and models applied to the description of electric phenomena in physics. No one would nevertheless consider that the science of electricity is "the" explanation of the complex functions of the nervous system. The parallel is not without historical insights: some biologists of the 19th century also considered that electricity was the explanation of life.

Another way to reach the same general conclusion on the future of biology is to compare the present situation with what happened in the 1930s when molecular biology arose. There are interesting similarities: the shared feeling that the plethora of observations should be replaced by the description of simple laws and rules; the movement of physicists and mathematicians – and today of bioinformaticists - towards biology, bringing with them new tools but also help in the search for simple rules that biologists, blinded by the accumulation of data, have not been able to distinguish (Morange 1998). This similarity pushes many to consider that biology is at the dawn of a new scientific revolution, comparable to the rise of molecular biology. The paradigm of molecular biology is dying, and ought to be replaced by a new one.

But this "Kuhnian" diagnosis omits to mention two main differences from what happened in the 1930s. The first difference is the ignorance at that time of the organization and properties of one domain of the world, extending between the molecules of the organic chemist and subcellular structures barely discernible under the light microscope. This was "the world of neglected dimensions" according to the physical chemist Wolfgang Ostwald, the world of colloids (Ostwald 1917). No such situation exists today, and the limits of present knowledge are of a very different nature. The second difference is the status attained by biology in the last half-century of its development. A considerable amount of knowledge has accumulated. New technologies have been designed which have hastened this continuous accumulation of data, but have also opened the way to an efficient manipulation of organisms.

Even the feeling that the "secret of life" remains unsolved, that life has been shot by molecular biologists without being understood, to paraphrase Stanley Shostak's words that the death of life was the legacy of molecular biology (Shostak 1998), is not the proof that something essential is missing. Maybe the romantic idea that the secret of life is hidden somewhere should finally be abandoned. Even the apparently scientific conviction that the laws of complexity will tell us something fundamental about the characteristics of living systems is probably also an illusion. Nothing essential is missing. What has to be done now to understand the characteristics of life, and the mechanisms of its emergence, is this slow process of interlinking different explanatory schemes. Perhaps some new explanatory schemes will emerge. But they will not replace the previous ones, but rather add a further level of complexity to this progressive interweaving of explanations. My conviction is that this diversity of explanatory schemes in biology will be irreducible in the future, as it is rooted in the very nature of the "objects" studied.

Chapter 3
Compromising Positions: The Minding of Matter

Susan Oyama

Many religions have understood language to be in some way primordial to the material constitution of the Universe
(Grassie 2005).

3.1 The Word

Au commencement était le Verbe.
Am Anfang war das Wort.
In the beginning was the Word.

The primal Word, bringer of order and meaning to chaos, introduces a comparison of Divine *Logos* with what I call *Biologos*. Both involve notions of direction, guiding agency, creative purpose, and meaning—roughly, intentionality. In addition, saying the "same thing" in several languages implies meanings that are independent of their linguistic vehicles, suggesting another characteristic of *Logos* and many information concepts in biology: their transcendence of, indeed, domination of the material.

The divine Word is classically contrasted with matter. Genetic information is said to be weightless and independent of its material substrate; for evolutionary biologist Williams (1992a) it dwells in a "codical domain" separate from the physical one.[1]

As argued elsewhere (Griesemer 2005; Griesemer and Wimsatt 1989; Moss 2003; Oyama 2000a) the concept of biological information carries underappreciated difficulties. The idea of heredity as information flow between generations, for instance, perpetuates the neoDarwinian neglect of development, bypassing the changing phenotype and positing instead the "transmission" of abstract instructions or programs. Thought to contain the information needed for the organism's makeup and functioning, these latter have a preformationist cast; they "explain" developmental processes that instead must be investigated in all their concrete complexity. Meanwhile, "genetic and cultural information," markers and spawn of all manner of nature-nurture distinctions, undermine efforts to integrate biology

S. Oyama (✉)
Professor Emeritus, Psychology Department, John Jay College, New York, NY, USA
e-mail: soyama48@verizon.net

[1] Later (1997, Chapter 9) he gives four: material, moral, mental and codical.

A. Barberousse et al. (eds.), *Mapping the Future of Biology*, Boston Studies in the Philosophy of Science 266, DOI 10.1007/978-1-4020-9636-5_3,

and culture. Through all this, and despite the difficulties of defining it coherently (Boniolo 2003a; Griffiths 2001; Godfrey-Smith 2007; Hayles 1999; Kay 2000; Sarkar 1996a,b), enthusiasm for information seems matched only by the oddity of its conceptual relations, including its suggestion of nonmaterial causation.

Here I explore only a small number of interwoven informational themes, first in two bodies of Christian writings on evolution, then in biology and philosophy of biology. I draw from the more infophilic members of each group, identifying positions by public stance, not private conviction. Christians and biologists are overlapping, heterogeneous populations; I will usually emphasize differences between groups and similarities within them (the major exception is their shared commitment to infotalk itself, with all its mentalistic connotations). By *biology*, I mean standard naturalistic science, contested as the phrase may be. I aim not to give a general history or an exhaustive analysis, but to suggest an intriguing degree of similarity among these religious and secular groups' treatments of *information*. Finally my concerns are with the ways we all think about biological processes, and with the interaction between science and the larger society. I will argue that although *Logos* and *Biologos* are often pitted against each other, they are intermingled in ways that complicate the increasingly noisy debates about science and religion in a growing number of countries, perhaps most visibly in the US. These minglings should push scientists to reconsider at least some of their infotalk, for reasons of both (inter)disciplinary coherence and workable relations with other audiences. Whether they make the religious hesitate is another matter.

3.2 *Logos, Biologos*: Four Themes

3.2.1 *Logos: Divine Information*

One of the bodies of Christian writing comes from the *Theistic Evolutionists* (TE), to follow physicist and theologian John Polkinghorne (2000, 76; *evolutionary theists* in John Haught 2003, 92), who accept the Darwinian narrative of gradual descent by natural processes and generally deny the special creation of species and other direct intervention by the deity.[2] Rather than seek evidence of God through science, like natural theology, TE sees evolution as enriching religious understanding. This active engagement with evolutionary theory contrasts with the separatism or rejection preached elsewhere.

The second set of religious writings is from the more visible theory of *Intelligent Design* (ID), a minimal repackaging of "Creation Science," which it follows in vehemently contesting Darwinism. ID seeks to overthrow scientific naturalism, claims to be a scientific program comparable to Darwinism, and demands a place in the science curriculum, using an aggressive mixture of public relations and litigation. It holds that unguided processes can't account for life's complexity, but usually

[2] Polkinghorne (2000, 97; 2003, 73) seems to want room for angels and miracles.

declines to name the designer.[3] For each informational theme I look first at TE, then ID.

(i) Source of Order: This theme has to do with the creation of living order, especially complex, hierarchical order. In Haught's 2000 (70–73) theological treatment information is variously metaphorical; not mystical and as real as atoms; and an active ordering principle. It exists as possibility until it uses ordinary physical causes "in its ordering activity." This is not the "stricter sense" of the genetic code, but Haught does use that code to argue for the irreducibility of information to chemistry: "God could be ... the ultimate source of the novel informational patterns available to evolution." Dismissing attempts to relate the deity to lower level efficient causes, he seems to invoke the Christian distinction between primary (divine) causes, by which God acts directly, and secondary (indirect, natural) causes; the latter, Haught holds, adequately explain evolution.

For Polkinghorne (2000, 72, 96, 147–149; 2003, 96), information is tied more to uncertainty, chaos and complexity, to absolute unpredictability and openness to the future; both he and theologian Holmes Rolston (1999, 20) cite Stuart Kauffman on the generation of order (see also Haught 2003, 67). Skeptical of primary and secondary causes, Polkinghorne (2000, 115–116, 124) does say that God acts through pure information, while creatures act through a "mixture of energetic and informational causalities."

According to Rolston "the really spectacular constructions ... manifest in biological diversity and complexity do not appear without the simultaneous genesis of information ... the key to all progress in biological nature" (1999, 14); it is a "locus of creativity" (357). Sometimes it is active: Polkinghorne speaks of "*holistic* ... top-down active information" and "pattern-forming informational causality" (2000, 100, 122; 2003). It "powerfully patterns subordinate natural elements and routines into hierarchically distinct domains" (Haught 2000, 70). In the genes, Rolston (1999, 39) affirms, "there is information superintending the causes; without it the organism would collapse into a sand heap."

ID, on the other hand, uses the natural theological argument from design: organisms are too improbably complex to have appeared without a designer. "Intelligent design is just the *Logos* theology of John's Gospel restated in the idiom of information theory," declares ID theorist William Dembski (1999; see also Johnson 1999). From this spare starting point he constructs an elaborate theoretical edifice, complete with a "Law of Conservation of Information," to show that complex design must be created by a prior intelligence (2001, 571; see Godfrey-Smith's 2001b critique). Similarly, biochemist Michael Behe (Seiglie 2005) insists that unguided natural processes are incapable of generating the requisite information for "irreducible complexity," which cannot evolve gradually because all parts must be in place for the whole to work at all.

---

[3] Some admit that ID is Christian (Johnson 1999). It includes believers in both "old Earth" and "young Earth." See Forrest and Gross (2004), Pennock (2001) on the infamous "Wedge" strategy; http://www2.ncseweb.org/wp/?p=80 has a link on the Kitzmiller case, in which evidence emerged about ID's direct descent from creationism.

Theistic Evolutionists stress information's elusiveness: it hides, is patient, even stealthy. It "slips silently out of our grasp" (Haught 2000, 77).[4] Rolston (1999, 24, 368) notes the difficulty of quantifying information, and says that "divine selection" wouldn't be visible. "Chance is an effective mask for the divine action. Still, God could be slipping information into the world," rather as though He were occasionally loading the dice.

Elusiveness is not an issue for ID; the detectability of at least "Complex Specified Information" (Dembski 2001) is the whole point. Less coy than his colleagues about origins, former law professor Phillip Johnson (Van Till and Johnson 1993) asserts that "unbiased scientific research" would find the method God used to create the world.

(ii) Counter to Chance: This is closely related to the first theme. The classic argument from design is that life couldn't have occurred "by chance." Although TE accepts Darwinism, its adherents sometimes say that evolution can't create without divine information. As Haught (2003, 87) observes, ID ignores "nature's own self-organizing spontaneity," but he joins many others in rejecting the materialist claim that "blind" selection will do the trick.[5] Rolston (1999, 355) asks where all that order came from, saying we can't shake a box of letters and get a poem. He continues that while the "know-how" for making salt is already in the elements, the information for an organism or a molecule of hemoglobin is not in the atoms. Such statements are common in arguments for *something* besides "chance" to account for life's complex structures.

In like manner, ID stresses the unthinkability of "random" processes or algorithms that could yield a cell or a plant (Dembski 2001).

(iii) Intentionality: Value, Purpose, Meaning: This is the least cohesive theme, partly because of the joining of planned action and meaning in intentionality. Consider the meshing of act and purpose in Polkinghorne's (2003, 67) remark that informational causation involves "agencies of an holistic and indeed, ultimately purposeful, kind."

Rolston (1999, 38–41, 359), calls genetic information the source of "intrinsic value." Talk about an organism's good is not just metaphoric for him: "the genetic set is a *normative* set, distinguishing between what *is* and what *ought to be.*" "Over evolutionary history, values have been generated," starting with survival. God is the "Ground of Information." For Haught (2000, 80), an informational hierarchy is meaning's bastion against relativism: "reality and human values have a sacred origin and a significance beyond what the physical sciences and biology can discern. Hierarchy is needed if some things are to be considered more valuable and more real than others." "The universe ... may have some overall meaning or 'point' to

---

[4] The Christian notion of *kenosis* is operative here (Polkinghorne 2000, 91–93). It is also consistent with TE's methodological naturalism that *Logos* not be scientifically detectable.

[5] Nick Matzke (pers. comm., May 5, 2006) holds that such statements refer not to any causal insufficiency of natural selection, but to certain evolutionists' use of Darwinism to proclaim a godless world. I return to this later.

it without ... needing to be displayed at the same level of investigation at which natural science functions (74)."

Insofar as ID has no official position on information's source, it should have nothing to say about this theme. We may, however, infer something from its mistrust of natural selection and its full-bore attack on materialism, relativism, and modernism.

(iv) Supplement to Matter, Energy, and Conventional Causation: Haught (2000, 61, 71) speaks of "the blank lifelessness of matter's physical simplicity" in the absence of the sacred. "Information, like mathematics, has a certain aspect of timelessness to it. It does not originate out of the historical past, nor does it undergo transformations in the same way that matter and energy do."

In addition to the matter and energy of physics, Rolston writes (1999, 14),[6] in a now-standard formula, "if we consult biology there is a third kind of thing: information." For Polkinghorne (2000, 96–97), it is not only biological: study of complex systems shows we must add to our store of causal principles "a third fundamental concept of a pattern-forming character that will embrace these emergent properties of holistic order.... It carries with it just a *glimmer* (no more) of the integration of the material with something that begins to look a little like the mental." We find "a hint of the emergence of the 'spiritual' in modern science" (163); God is pure spirit, pure information. The "passive information storage" of information theory requires energy to be read, but the active information through which God acts is "not mixed with energetic causality" (2000, 123–125; 2003, 61).

We find similar expressions in ID. According to Dembski (2001, 569, citing Monod) natural causes (chance and necessity) "in principle" can't produce "Complex Specified Information." Meanwhile, Behe says (Seiglie 2005) that "information is not matter or energy.... there is something else in DNA, and that is the intelligence component." Dualism is necessary because chemistry doesn't specify a sequence of nucleotides, just as the chemistry of book materials doesn't specify the text of a book (Johnson 2001, 545).

3.2.2 Biologos: Genetic Information

In biology, information is usually construed intentionally or semantically, not statistically, as in information theory.[7] Intentionality itself has a semantic sense in philosophy ("aboutness") and a vernacular one of planned action. I suggest that the joining of meaning and activity in many uses of *information* presents complementary problems for TE and for certain infophiles whose writings on evolution feature an explicit or strongly implied atheism (here called *Atheistic Evolutionists* or AE):

---

6 See Grassie (2005) on information's paradoxical scientific status.

7 Sterelny and Griffiths (1999, Chapter 5) *intentional* and *causal* information index much the same contrast, although *causal* is not ideal for Shannon information, which was about correlation and predictability. See Godfrey-Smith (2007).

TE's methodological naturalism militates against divine action in evolution, while AE's traditional materialism is at odds with mentalistic rhetoric. By embracing an active creative intelligence ID avoids at least this snag.

Intentional information's quality of aboutness involves *mental content* or *meaning*. It is operative in talk of genetic blueprints, programs, and other kinds of putative representations in the DNA. The possibility of *mis*representation or error is its hallmark: if nature is encoded in the genes, some developmental outcomes are not what they should be, but deviations from, even perversions of, preexisting representations. The political and moral mischief fomented by infotalk is usually connected to this semantic, intentional reading, often with tacit invocations of genetic essences (Oyama 2000b, 2002). Information theory, by contrast, is about the statistics of uncertainty, not meaning. The mathematical rigor of communications-engineering is what makes *information* both enticing to biologists and hard to appropriate.

With the advent of the "genetic code," information largely supplanted the older concept of *specificity*,[8] which referred to three-dimensional entities and their relations and carried less vexing (to some, including me) metaphysical baggage. When communications terminology invaded biology even the most critical skeptics, some of them later credited with shaping the informational age,[9] largely adopted what I've elsewhere called *the language of language* and here have dubbed *Biologos*. It eventually included *codes, messages, transcription, translation, editing, sense, nonsense*, and more. Specificity was no match for information, whose cachet flowed from physics, engineering, cybernetics, mathematics, wartime cryptanalysis and weapons control, not to mention the enthusiastic efforts of linguist Roman Jakobson; complexity studies and computer and cognitive science are more recent tributaries (Hammond 2003; Kay 2000; Keller 2000; Morange 1998; Moss, 2003; Sarkar 1996a). With the influx of newly interested physicists, biology ceased to be a "sissy" field.[10]

The present use of informational concepts in biology is variable. Though not universal, it is so well established it is generally deemed to need no definition, let alone justification. AE John Maynard Smith (2000), however, mounts a spirited defense of both probabilistic and semantic uses in biology (Godfrey-Smith 2000a, and Griffiths 2001, reply).

I suggest that *information*'s rich web of mentalistic connotations and its apparently uncontroversial scientific usage allowed its easy adoption for religion. More

---

[8] Not to be confused with Dembski's (and Dawkins'—see below) "prespecified information."

In the 1940s biologists spoke of the specificity of genes and their products, enzymes and substrates, antigens and antibodies, bacteria or viruses and their hosts (Kay 2000, Chapter 2). Morange (1998, Chapters 1–2) dates biological specificity to the late 1800s, charting its path "from a biological concept into a stereochemical one" in the late 1930s (14). He argues that physicists detached it from direct biochemical interactions to reduce it to physics and chemistry. See Keller (2000, Chapter 4).

[9] Kay (2000, Chapter 1) names notable dissenters (Chargaff, Cherry, Shannon, von Foerster), but they do not seem to have stemmed the tide. She notes that *information* lost its quotation marks in the '40s.

[10] Schrödinger's associate Neville Symonds (Kay 2000, 61); see also Keller 1990.

contentiously, perhaps, I maintain that the atheists and theists like information for much the same reasons, not least of which is that it seems scientifically impeccable and is yet somewhat mysterious.

(i) Source of Order: Genetic information is widely seen as the source of organismic order. Jacques Monod (1971, 14, 45) moved from rejecting teleology to embracing "teleonomic information" "*oriented, coherent, and constructive* activity."[11] Protein structure hides "the 'secret' to those cognitive properties thanks to which, like Maxwell's demons, they animate and build living systems" (81).[12] By the end of the 1950s, in fact, Monod lamented his earlier subordination of finalism to chance (Kay 2000, Chapter 5). Morange (1998, 78) makes the fascinating claim that an early-twentieth century Viennese preoccupation with "memory and ... a deep, hidden order" is present in Schrödinger's desire "to find a principle of order in organisms and link it to the memory of things past." François Jacob asserts, "Heredity becomes the transfer of a message repeated from one generation to the next"; he quips that computer programs "made an honest woman of teleology" (1975, 254, 9).

Like *Logos,* then, *Biologos* has a semantic, representional aspect, usually a description of the organism or a guide for building one. Insofar as they are prescriptive, such representations also play a somewhat agentic role, especially in accounts of development. Williams writes (1992b, 10), "genes direct development." A textbook tells us, in a new section on information (Arms and Camp 1995, 184),[13] that genes are "units of genetic information," the "inherited instructions for building proteins and other molecules, including *when* ... and *which kind* of cell should make it."[14] This information is the prime mover (Campbell 1982; Loewenstein 1999). It is also said to *maintain* organismic form (Solomon et al. 1996, 2), though the vividness of TE Rolston's organisms collapsing into heaps without it is hard to beat.

For infophilic Atheistic Evolutionists, *Biologos* is not elusive the way *Logos* is for Theistic Evolutionists. Champions of the argument from design, ID and AE alike, are convinced that it stares you in the face. It is one thing for TE to emphasize the intangibility of information: they accept both God and Darwin, so divine information cannot be an observable intrusion into the natural order. One would think, though, that information's shiftiness would give pause to the scientists supplying

[11] From Pittendrigh's (1958) *teleonomy*, or biological goal-direction. This drama played out against a rich socio-political backdrop in the 1950s, with notions of final causes and vitalism still influential among French intellectuals; Teilhard was significant, as was the Lysenko affair, over which Monod broke with the Communists (Morange 1998, 152; Kay 2000).

[12] Notice the minded molecules. The English title of Szilard's (1929/1964) paper on the demon was "On the decrease of entropy in a thermodynamic system by the intervention of intelligent beings."

[13] See Bresler (1971) for an earlier example; Loewenstein (1999, xvi) on this "cosmic principle of organization and order" in biology.

[14] See my (2000b) "cognitive-causal gene"; Moss (2003, 2). Kay (2000, 72) writes of genetic codes as "mediums and agents of information storage and transfer," while Keller (2000, 46–47) notes, "the gene was ... a locus of causal agency."

that account.[15] At the end of a 1956 symposium on information theory and biology, an uneasy consensus was reached (Kay 2000, 126–127). As a metaphor information had to be used with care: it resisted quantification and was always relative. Yet the "discursive potency" of the term was too great for these cautionary sentiments to have much impact.

Insofar as it remains unquantified, undefined or inconsistently used while still being widely accepted, then, information could be termed "elusive." There is always slippage in our words, and for those seeking to harness information's rhetorical power, such ambiguity and undercharacterization could be an advantage.

(ii) Counter to Chance: It is now common to refer to organismic organization or adaptive fit as design, without scare quotes. As in the theological discourses, information is said to counter "randomness" or "mere chance," and like ID, Atheistic Evolutionists Dawkins (1986) and Dennett (1995) anchor their accounts in the argument from design: certain highly complex objects, including live ones, could not have come about by accident. The designer in the AE version, of course, is not God but "Mother Nature," or natural selection (Note 40): information again keeps chance at bay. Giving a twist to this reasoning, Dawkins (1986, 9) asserts that there are fewer ways to be alive than dead (or not alive).[16] He defines complexity, or complicatedness, as "some quality, specifiable in advance, that is highly unlikely to have been acquired by random chance alone."[17] He disclaims any special attachment to his definition, requiring only that a substitute capture the difference between living things and "the objects of physics" (1986, 15), but chance seems central: he adds an amusing bit about the likelihood of getting a complex organ by haphazardly throwing cells about.

I find the notion of prespecification obscure, both in ID and here (specified by whom?), and wonder whether Dawkins' call for an alternative betrays doubt about its viability. Among other things, it would seem to assume that adaptive requirements of a niche exist independently of and prior to the organism (like evolutionary psychologists' "task descriptions," below)—indeed, since we are talking of organismic "design," perhaps even some unique set of solutions to those requirements.

(iii) Intentionality: Value, Purpose, Meaning: The kind of purpose that Jacob and Monod embraced was about biochemical self-regulation, not personal ends. Nevertheless, links exist between cybernetic goal-seeking and vernacular notions of mentality and purpose. The standard scientific view is that evolution has no meaning or goal, but that it instills goals or even meaning(s) in its creatures: roughly (in more senses than one), survival and reproduction. "The intention of a psyche has been replaced by the translation of a message," wrote Jacob (1973, 2).

---

[15] Part of the growing critical literature on information is cited in these pages. My impression is that infotalk still thrives in scientific practice.

[16] It is hard to say just what this means.

[17] Dembski's (2001) strikingly similar treatment cites Dawkins not as source but only as confirmation.

Dennett (1995, 325) exhorts us always to ask, *cui bono*? Who benefits? From the gene's eye view associated with him, Williams (1966), Dawkins (1976), and others, the entities whose benefit most often (or always) counts are genes, or more precisely, the abstract information housed in the DNA. Williams (1992a, 11–15) insists that objects don't have meaning, information does. For him, genes are not things but bundles of information; DNA is but a medium.

The practice of looking for reproductive benefit is not limited to the genetic level; focus on organisms and groups is historically older and persists today. Whatever the level, it is genetic frequencies that paradigmatically define evolution's bottom line.

Certain meanings and goals, then, are traced to natural selection and the genetic information it produces. There may also be an assumption that these define what is most real or important. In the era of expansive sociobiological explanation, biologists studied function by looking for current fitness maximization, reading the evolutionary reproductive "goal" directly into the regulation of ongoing behavior. They thus slid from a historical discourse of natural selection to an immediate causal one of motives and intentions (from "ultimate" to "proximate" explanation).[18] This led to talk of the *real* (selfish) reasons underlying behavior that was only *apparently* altruistic, nurturant, moral, etc. (Oyama 2000a, Chapter 9). Daly and Wilson (1988, 7) denied that they were explaining motivation, but still detailed the necessity of feigning sincerity while manipulating others to one's own advantage. They denounced "mystico-religious bafflegab about atonement and penance and divine justice" as really just "a mundane, pragmatic matter" of controlling others (255–256), and Alexander (1979, 133) saw the conscience as the "still small voice that tells us ... how we can cheat socially without getting caught." Such theorists' radical deflation of social relations in general and of human morality in particular provoked a fair amount of outrage both inside and outside academia. Their efforts to quell fears and clarify issues were apt to be subverted by their own presentational choices (Oyama 2000b, Chapter 6), a common enough problem today.

Sociobiology's successor, evolutionary psychology, has managed to reduce these particular problems by talking about evolved mental mechanisms, rather than bypassing the psychological as their predecessors had been apt to do. Still, much of the same theoretical apparatus remains in place. These theorists draw inspiration from the information-processing models of cognitive science, appealing to a universal human "genetic architecture," shaped by natural selection, which defines the

---

[18] In actual usage, Mayr's (1961) distinction can resemble the theological one mentioned earlier, between primary and secondary causes, with the attendant uncertainty about the precise relationship between them. The slippage between the historical-causal and the quasi-theological explains much of TE's objection to evolutionary theory, and AEs exacerbate matters when they point to their technical definitions but nevertheless issue evolutionarily ultimate explanations that *sound* theological. We cannot entirely banish metaphysical projection from our science or from any other activity; all the more reason to mind what we are doing (it matters!) and to face up to it. See B.H. Smith (in press) on public and private ontologizing and Winther (2006) on the need for dialogue.

tasks the mind must perform (Tooby and Cosmides 1992). Our goals are "built in by evolution" (Barkow 1991, 297).[19] Meanwhile, teleosemanticists account for meaning and intentionality by pointing to a history of natural selection.[20]

(iv) Supplement to Matter, Energy, and Conventional Causation: In biology's informational age nature is interpreted as matter, energy, and information (Campbell 1982, 15–16; Jacob 1973, 95).[21] Katherine Hayles (1999, 51, 62) has chronicled the disembodiment of information during the early synergistic growth of the information sciences and molecular biology. "The triumph of information over materiality was a major theme" at the first Macy conference on cybernetics, where John von Neumann and Norbert Wiener stressed that information, not energy, linked humans and machines. Warren McCulloch saw the inconsistency of speaking of disembodied information in the nervous system, noting that information flows via signals, and signals require embodied existence. Such scruples notwithstanding, the trend was firmly toward the discarnate.[22] Decades later, Dawkins (1995, 19) announces, "life is just bytes and bytes and bytes of digital information."

3.3 Serving Two Masters: Minding Matter

3.3.1 Dematerializing Matter

Information serves two masters, it seems, and divided loyalties call for vigilance. It's not just that scientific biology is historically continuous with natural theology, or that religion and science borrow from each other, or even that a person can be both theist and scientist. Robert Young (1985) describes the turmoil attending early evolutionary theorizing and the extraordinary difficulty of communicating

---

[19] See Stotz and Griffiths (2001); Sterelny and Griffiths (1999).

[20] Dennett (1995); Griffiths (2001); Millikan (1984); Sterelny et al. (1996). Maynard Smith (2000) employs the same strategy, but not the term. Teleosemantics is not uniformly infophilic. Sterelny and Griffiths (1999, 105) say the "apparently magical nature of intentional information is one of the major objections to a materialist account of thought," describing teleosemantics as an attempt to neutralize the magic.

[21] Biochemistry "went informational" in the 1960s (Kay 2000, 13). "A noncorporeal quality inherent in the intelligence of the programmed system, biological information served to demarcate the RNA ... from mere brute matter" (231). An early '70s text: "Life processes require an unbroken flow of energy, matter, and information. The flow of energy and matter is completely determined by the function of various proteins. ... Information flow is another matter" (Bresler 1971, xiv). See also Solomon et al. (1996, 2). Morange (1998, Chapter 15) adds shadow and nuance to the usual triumphalist account of molecularism's spread.

[22] For Wiener it was not matter but "the memory of the form" that was passed on in cell division and genetic transmission (Kay 2000, 84–89; Heims 1991 details the discussions of information at the Macy conferences).

effectively about it. In Darwin's time, as in ours, there was anxiety over the possible replacement of a theological explanation of life by a natural one.

The quotes in the last section don't, presumably, signal allegiance to metaphysical dualism. Indeed, infotalk is sometimes seen as dualism's final repudiation: Dennett (1995, 195) hails certain models of molecular evolution as "Darwinism triumphant, reductionism triumphant, mechanism triumphant, materialism triumphant," and just before his passage about life as bytes, Dawkins (1995, 18) ridicules vitalism, proclaiming, "there is no spirit-driven life force, no throbbing, heaving, pullulating, protoplasmic, mystic jelly."[23] Yet perhaps one can understand a certain degree of confusion, or perhaps, strategic exploitation. ID's Johnson (2001, 544) pounced at the seeming dualism of Williams' nonmaterial codical domain, mentioned earlier.[24] To defend Williams, Dawkins (2001, 550) cited substrate neutrality—the idea that a pattern can be instantiated in a variety of materials—insisting that dualism didn't enter at all.

Indeed, I think there are (at least) two related ideas feeding this desubstantialization of information. Without necessarily entailing dualism, each communicates a somewhat equivocal attitude toward matter. The ideas are that (1) certain kinds of organized matter cannot be accounted for by traditional (physical) methods and (2) information is independent of substrate (think of the translations that began this essay).

On the first, recall the evolutionary argument from design, or consider Maynard Smith (2000), who adverts to Monod's concept of *gratuité*—generally interpreted as the arbitrariness of certain biochemical relations, say between a gene and its repressor—to justify calling genes *symbols*.[25] One commonly finds the claim that the laws of chemistry don't *require* just these correspondences, so some additional ordering principle is needed.[26]

Information's substrate neutrality, or independence of medium, is central to Dennett's *Darwin's Dangerous Idea* (1995), while in the field of Artificial Life,

---

[23] I can't help smiling at the gross bodily dampness of his imagery.

[24] This domain inspired Haught (2000) to adopt the language of information. It is much less evident in his 2003 book, owing partly to attempts to use it to assimilate him to ID (Behe 1999). Haught now prefers a more purely metaphorical usage of the term (pers. comm., Nov. 22, 2005).

[25] Boniolo (2003b) argues that *gratuité* carried intentional existential connotations, but that *contingent* is a better translation than *arbitrary* or *symbolic*.

[26] This seems an odd elevation of time, space, and organization to the transcendental. There was an apparently analogous impulse early in the twentieth century, among British emergentists who distinguished chemistry from physics by a non-physical ordering principle. Table salt's properties, they said, were not explained by Na and Cl alone; some principle beyond the laws of physics was needed (Mohan Matthen pers. comm., Aug. 26, 2006). Recall Rolston on salt and Johnson on the chemistry of books. Nick Matzke (pers. comm., May 5, 2006) reports that ID inserted this reasoning into the Kansas school science standards on evolution, which continue to be whipsawed by successive elections.

Emergent properties and contingent specificities are common: does the failure of the "laws of physics" to dictate the precise disposition of minerals in a rock or the shape of a stream bed call for an addition to matter and energy? Cf. Godfrey-Smith (2007).

the living is defined by form: abstract pattern, not material organisms (Hayles 1999, 231).[27]

TE, ID, and AE seem to agree that materialism is at stake in the conflict over evolution. Historically, active matter has threatened God's dominion, but the forced choice between inertness and agency (or passivity and activity), like the one between matter and form, is a trap in which dead matter requires an active principle to move it, be it God or information.[28] In 1982 Ernst Mayr depicted natural selection as the updated hand of God, and Kay (2000, 163) finds ample historical precedent for seeing the Book of Life as "scriptural representations." She asks whether there might be "tacit theistic investments in the [genetic] sequence as the primal word of genesis" (see also Moss, 2003). My question is, just how much of this package do we really want to retain?

3.3.2 Compromising Positions

This part of this paper sketches out several issues that are implicated in our themes, and that I think are problematic for scientists. They are *chance, essentialism*, and what I'll call *compromised materialism.* Their difficulties are amplified by infotalk, and as was the case with our themes, they are tied together by the *minding of matter,* understood both as *instilling the material with (nonmaterial) intentionality* and as *paying attention to, being careful of.*

(i) Chance: Après-Monod, chance and necessity are nearly synonymous with evolution.[29] But these terms are slippery. Chance means both "random" and "not intended,"[30] as in "We met by chance"; it often hinges on uncertainty, not anything

---

[27] Sterelny and Griffiths (1999, 361–363) criticize the matter-form dichotomy for eliding multiple levels of abstraction.

The ability to be manifested in multiple media is not the same as being independent of medium: there are quite exacting material requirements for a computer program, a base sequence, or a printed word on a page, with sometimes dire consequences when those conditions are not met. Boniolo (2003a) highlights just such uncertain conditions. As responsible people know, the reliable always risks invisibility.

[28] Reviewers of this paper asked whether my position is materialist. In some sense it clearly is, but simply to assent, without a great deal of explication and qualification, would be to ratify a historically complicated set of contrasts and assumptions that is better disputed than affirmed. This is not the place to discuss either metaphysics in general or mine in particular, but I suspect my (God-free) ontology leaks from every page I compose. The curious can consult any of my work, where passive, formless matter is not contrasted with active informing mind (God's or Mother Nature's). Matter needs neither vital forces nor their modern envoy, information, to *inter*act in the most astonishing ways.

[29] For Monod (1971, 118) natural selection feeds on "the products of chance," while Dennett (1995, 520) attributes "design work" to chance and necessity. Tooby and Cosmides (1992, 63) say "chance and selection [are] the two components of the evolutionary process ... all aspects of design must be attributed to one of these two forces." See Boniolo (2003b); Godfrey-Smith (2001b); Stotz and Griffiths (2001).

[30] And much more. Boniolo (2003b) gives a typology. Dawkins (2005) actually contrasts adaptations with "the unintended workings of physics."

specific about underlying causal processes. Believers in Intelligent Design say design must be *added to* chance and necessity, while many biologists see the pair as *producing* design: ID wishes to invoke divine power, AE to exclude it. Dembski (2001) denies that information can be generated by natural laws or algorithms, while for Dennett, natural selection *is* the algorithm that produces biological information. Chance in these exchanges generally seems to mean something like "everything but the force I have in mind." Actual randomness is rarely at issue. Mutations are not equiprobable, and neither are phenotypic variations. Instead, these oppositions function to signal that the chosen explanation is the only one thinkable; everything else is rendered unintelligible by the label, "chance." (If you don't accept my explanation—God or natural selection—you must think that agitating a box of letters will give you a sonnet or that cells tossed in the air will come down as a kidney.) It hardly helps when AE uses this opaque notion of chance *both* as a crucial component in its causal account (chance and necessity) and as the absurd alternative to that account (getting an organism "by chance").

Using *chance* as a cover term for (some or all) *natural processes*, furthermore, creates perpetual uncertainty about whether it is being used as a contrast to a transcendental story or a causal one. To the religious writers, it seems to mean *not intended, known, meant, or given meaning by an intelligent God.* If so, the challenge for TE would seem to be to separate these from the action component of *intention*, not easy given the intimacy of the connection (in contrast with, say, *wish* or *want.*) And what does *chance* signify for AE? Why is natural selection its antithesis? If information for the Christian writers is a vehicle for *Logos*, what does it channel for AE? Why are readers encouraged by AE to read selection intentionally, and then scolded when they do so? Darwinism's official line, after all, is that Nature is not intelligent, does not plan or have meaning.

Pitting *Biologos* against chance, like opposing *Logos* to chaos, promotes it to a formative principle while implying a radically disordered world. Dawkins (1986, 128) says, "a dead planet has atoms, molecules and larger lumps of matter, jostling and nestling against each other at random, according to the laws of physics." Inert, "dead" matter needs an active formative principle; hence, information and/or God. Yet what natural selection needs, and has, is precisely a *historically ordered world*—physically, chemically, geologically, climatologically, biologically. For some kinds of organisms consistently to reproduce more than others in a way that eventually shifts the population, they must reliably develop into the same sort of being, and live their lives in more or less consistent relations with their surroundings. Hence those surroundings must themselves be fairly ordered, and indeed they are, partially as a result of past interactions, many with those very organisms and their ancestors, as well as other organisms. Even a world without life would still have the first four of that list (physical through climatological). And I was absent the day when "according to physical laws" became "random."

Casting evolution in terms of chance and necessity (and then contrasting evolution with "chance") not only scrambles the mind; it conceals *worlds* of multileveled (causal) complexity. Where, for instance, would one fit discussion of the developmental processes without which there would be nothing to select?

Both of the religious groups under discussion (as well as some nonreligious evolutionists) dislike what they see as AE's use of evolutionary theory to advance a traditionally materialist (atheistic) metaphysics. But ID also protests aspects of science's causal account, whereas TE claims not to. If the whole vocabulary of information and chance in this literature is riven by the cause-mind ambiguity in the ways I've suggested, TE's denial of the "blindness" or "unguidedness" of evolution is misplaced and strategically unwise, for it appears to challenge Darwinism's explanation by natural selection, when it really objects to AE's promotion of a godless metaphysics (see Note 5). Add its occasional suggestions of divine infusions of information into evolutionary processes and the stage is set for crossfire and uproar.

On some important issues TE and many secular evolutionists are allies,[31] but if the former are perceived as equivocating on the (causal) unguidedness of evolution,[32] an effective joining of forces seems improbable, and when evolutionists employ a vocabulary suffused with intentionality and insist on deploying a suspiciously bivocal notion of chance, they roil already murky waters. Information not only serves two masters, it is itself (at minimum) two-faced. In it, the semantic and the causal (plan and act, word and deed) are joined, as they are in the notion of intentionality (see Note 14). If TE really accepts a naturalistic account of evolution and objects only to AE's attempts to drain meaning from the world, and if AE really insists only on that causal account and does not wish to pronounce on the possibility of anything outside the naturalistic framework, they had better say so. Infotalk is of doubtful utility in this project.

(ii) Essence: Although essentialism was supposed to have been banished from biology long ago, it endures (Oyama 2002). Essences in the Christian writings do not surprise, but they are an aspect of *Biologos* as well. Species and individual natures are said to be carried in the genetic information; witness evolutionary psychology's universal "monomorphic mind" (Barkow et al. 1992). We have essence as soul and character as well, though not always as openly (Kay 2000, 53; Keller 2000, 47; Nelkin and Lindee 1995; Oyama 2000b). The popular press is rife with references to the DNA of soldiers, Hollywood, corporations—any entity whose most fundamental nature is being characterized. Genetic essence feeds fantasies of duplicating ourselves indefinitely. In one magazine article cloning was conceived(!) as hi-tech reincarnation. More consequentially, essences carried in the DNA support the belief that a full person is present from the moment of conception (ensoulment), so that stem cell research is an abomination and abortion, or the discarding of a fertility treatment's unused embryos, murder.[33]

Given genetic information's agentic and representational functions, such natures are both causal and normative, speaking to notions of normality and health, even monstrosity (Oyama 2000a, Chapters 7, 9, 10; 2000b, Chapter 5; Wolfe 2005).

---

[31] Haught testified against ID in the Kitzmiller case on evolution in the schools (Matzke pers. comm. May 5, 2006).

[32] That is, by God. I leave aside the question whether mutations are completely orthogonal to selection.

[33] It is a small step to charges of criminal neglect or homicide for imprudence during pregnancy.

Intentionality carries the possibility of misrepresentation, so an organism can harbor, or even *be* a mistake. This gives a particular spin on human variation and pathology. If genetic information represents us and our goals, then we can be wrong in our very essence (if our DNA is disordered) or untrue to it (if we pursue, say, a disordered "lifestyle"). It seems that essentialism, like teleology in Jacob's witticism, had her virtue restored by infotalk.[34]

(iii) Compromised Materialism: What's the Matter With Matter?: Infotalk often signals a flight from materiality. With von Neumann's self-reproducing automata, "genes became essentially an 'information tape'" (Kay 2000, 111). Hence Williams' definition (1992a, 11) of a gene: not a molecule, but "the transcribable information coded by the molecule." Or, for the more poetically inclined, here is Dawkins (1995, 4), conjuring the image of a "river of DNA . . . it flows through time, not space. It is a river of information, not a river of bones and tissues: a river of abstract instructions for building bodies, not a river of solid bodies themselves. The information passes through bodies and affects them, but it is not affected by them on its way through."

Genetic information was born in the embodied, stereospecific relations so stunningly explored by Jacob and Monod years ago, so the centrality of substrate neutrality for infophiles like Dawkins and Dennett is piquant. Forsaking wet, bumpy little lumps for the empyrean abstract, they demote molecules to *mere stuff*, while designating only some of them as vehicles for *Biologos*. Another aspect of minded matter, then, is privileging, setting off certain causes as more formative than others: some things "carry" information, while others do not, even though all objects interact selectively, and the differences and correlations that support information theory are ubiquitous. We can't do without abstraction, and the yearning to leave messy materiality behind needn't be pernicious; the danger is that the tether can snap, letting the world of objects fade from view. Some of that world's features, excluded from the formalisms, become invisible, and others, regarded as something other than primary, may be wrongly neglected (Moss, 2003; Oyama 2000b). Keller (2000, 79) observes that the concept of gene regulation eclipsed "any notion that genes might rely on nongenetic factors for instructions."

Yet researchers managed before the information age. The older concept of specificity implied *material interaction* rather than disembodied agency or formal code, and so evaded the "inert matter vs. active mind" snare. Both specificity and information "signified the complementarity of highly ordered biological structures," but the former was a three-dimensional Aristotelian material cause, whereas information was formal. "Specificity corresponded to body, information to soul" (Kay 2000, 41).[35] "Because molecular specificity was immobile and grounded in matter, information came to serve as its carrier beyond material bounds.... Possessing motion, information could transcend the limits of structure. Specificity was mute; information communicated specificity's messages" (53). Sahotra Sarkar

---

[34] See Amundson (2005) on essentialism in evolutionary studies.

[35] The preference for the abstract over the material extended, Kay (2000, 269) argues, to molecular geneticists' tendency to favor reason over test tubes.

(1996a) ends his sharp critique of the information concept by suggesting a return to *specificity*. Whether or not one heeds his call, or agrees with him (1996b) that infotalk has inhibited study of large biomolecules as physical objects, any term makes some things more salient than others, and this can be consequential.

I appreciate *specificity*'s connections to material organization, but doubt it makes sense to substitute it for *information*, even if such things were legislatable (but see Morange 1998, 152, on adaptive vs. inducible enzymes). The sprawlingly heterogeneous usage of *information* has only been touched on here, and part of my point is that the search for a one-size-fits-all organizational principle is misconceived. The same could be said for other terms that have local utility for particular interactions but questionable value when made too general (up- or down-regulation of synthesis rate, say, versus "regulation of development").[36] I cannot adjudicate *specificity*'s adequacy for biology's many specialized areas of inquiry, whose number and technological sophistication have vastly increased since the informational revolution. In certain contexts it might do, but for many purposes it would surely need additional ... specificity.[37] How much and what kind depend on the task at hand; this search for appropriate conceptual breadth, however, is endemic to any use of language, scientific or not. I do think it would be quite a salutary practice to ask what work, if any, is being done by *information* when it is used, then to look for phrasing that is less ambiguous and laden with extraneous implications. This will often mean staying rather closer to the phenomenon, whatever it is. It might be a correlation, a predictable developmental reconstruction, the appearance of one pathway rather than another, even a bit of behavior that seems to appear without training. (Something similar exists for terms like "innate." It is not that they never have reasonable referents, but that even when they do, these are illegitimately cross-referenced with quite different meanings.) I suspect that such rigorous questioning will sometimes reveal that no work is being done at all, or that what is being done can't be endorsed in good faith. It might amount to a gesture toward technical mastery—maybe even the eventual discovery of the master program, a promissory note I would not sign—or less innocently, a covert, unprincipled way of privileging some causes over others. This coming up emptyhanded would be instructive in itself.[38]

In any case, TE and ID should have less faith in scientific infotalk. I intend their enthusiasm for information to function as a mirror for biologists. Perhaps, seeing

---

[36] "Biological agency," says Moss (2003, 72), "is rediscovered in *process* and at many levels of context." The unitary regulating agent gives way to "regulation" by systemic interaction. In biology, as in society, we should ask, who/what regulates the regulator?

[37] Kay (2000, 45) herself notes the term's "metaphorical" and "heuristic" qualities and its complicated relations with information. The goal is not to find pure, nonmetaphorical terms, but to notice whether and how they work. We can make "empirically accountable, problem-solving-oriented claims and still make reflectively explicit the rhetorical resources that are mobilized in doing so" (Moss 2003, 73).

[38] An hour in the library turned up numerous biology texts that invoked information (-flow, -processing, etc.) in the first few pages but did not put it in the index.

what looks like a beam in the eye of the faithful, nonreligious scientists will notice a mote or two in their own.

TE and ID want biologists to forsake materialism (metaphysical in the first case, metaphysical and methodological in the second). I suggest that when AEs deify information they seem to lack the courage of their materialist convictions. This invites misconstrual and misuse. Rhetorically making matter disappear also encourages neglect of the histories and concrete arrangements—time and space—that generate biological marvels, and makes it harder to communicate them effectively.

Why this devotion to nonstuff, finally, to this nonmaterial formative principle with its theistic overtones, its ambiguous relations to ordinary causality and "substrates," its recalcitrance with respect to quantification?[39] Part of *minding matter* (in my second sense, of being vigilant) is remembering that abstractions are *from* something. If we suppress spatial and temporal relationships, say, *and forget that we've done so* (the tether snaps), and if we additionally collapse the levels between observation and formalism, we can end up believing that humble lower level to be incapable of generating what we've seen, and call in the formalism to do it instead. Those bumpy lumps have an embodied history that helps explain their present interactions, as do the immediate surround and ambient conditions in which they are embedded. Scientists presumably know this, but sometimes it's hard to tell.

3.4 Conclusion

Serving two masters means keeping an eye out in more than one direction at once (being cockeyed?). My main concern here is with talk among scientists and their connections with the wider world. Young (1985) tells how the anthropomorphism of natural selection both helped and hindered Darwin's efforts to convey his ideas. While it made evolution accessible by drawing on established beliefs about divine creation, it also confused the conceptual scene for just the same reason.[40] If most scientists, a century and a half later, are intent on persuading the public that evolution is not intelligently guided, that organismic development, structure and function can

[39] See Maynard Smith (2000) for an ingenious try. While professing faith in the uncontroversial nature of coded information in the DNA, he confesses that it is not well understood how the DNA and proteins program development. Notice that his doubt applies only to *how* the program works, not its intelligibility as a concept.

Sampson's book (1984, 4–5) focuses on "biological systems capable of autonomous execution of stored programs." In a section on "developmental programs" (139) he notes that these could involve an "immutable specification of the program coded in the genome itself" or, as seems to be the case, sequences of "causally linked events." I wonder, in the second case, what/where is the stored program? But how likely is it that this uncomfortable question would be addressed in a book named *Biological Information Processing*?

[40] Godfrey-Smith's (1998) criticism of Dennett's use of "Mother Nature" can be extended to the minding of matter in general: first loading an account with intentional terms, then constantly trying to correct readers' understandable "misunderstandings."

be explained naturalistically, why would they persist with a vocabulary so replete with mindedness?

I've mentioned some of the historical factors that fed information's juggernaut path to dominance, and I don't believe that scientists should or could undertake a massive purge, even if, as Keller (2000, 63) points out, in recent years "the gene has lost a good deal of both its specificity and its agency." Particular information concepts have done work. This is not to say that they were always, or ever, the best choice, and I do think that more reflectiveness is in order; when ways of speaking become naturalized their implications tend to pass beneath the radar. Perhaps such reflection would lead to less sweeping and grandiose usage, to more scrupulous definition and restriction, to more disciplined (and potentially theoretically fruitful) attention to phenomena left out of those snappy formulas. A reduction in purple prose seems a small price to pay for such benefits.

But perhaps those who are most prominently engaged in the public debate, like Dennett and Dawkins, don't really *want* to distinguish cleanly between naturalistic and theistic accounts. Perhaps the strategy is to exploit the ambiguity Young spoke of by aggressively taking a position on the opponent's turf, saying, as US general William Boykin did of a Muslim antagonist, "I knew my God was bigger than his" (Los Angeles Times, Oct. 16, 2003). (Indeed, a review of Dennett's new book on religion was graced with a cartoon of God sending a lightning bolt out of the sky at a man, and of the man hurling back a double helix.) I am not convinced that this is the best way to go about dealing with the issue.

Many may find the current polemics in the US too bizarre and distant to be of much concern, and the evolution debates are not the most disturbing things that are happening there. Yet this is not just a struggle over biological theory in one country with a largely Protestant citizenry. Over the last few decades the landscape has changed, as landscapes will, partly due to our own activities. Evolutionists across the globe are increasingly in direct confrontation with religious fundamentalists from a variety of traditions, before ever larger audiences. The Vatican seems conflicted about undirected evolution, whatever this may mean, and even the science-friendly Dalai Lama is doubtful (*New York Times*, Oct. 2, 2005). Meanwhile, the internet is full of ferment over Islam and evolution.

I think that Justin Smith (2005) is correct in identifying the current struggles over Darwinism as part of a more general disquiet about modernism. This last is pervasive indeed, and fervent fundamentalism comes in many flavors. No one knows how to steer these treacherous waters. Nor can we control the uptake of our work, but we can surely be mindful of our practices, including our speech to ourselves and to others. We might ask whether we have full confidence in the ability of energy and matter to produce the living world (absent outside agitation, as it were). We might remember the "brute matter" that is the basis of our science. Lowly as it is, if granted a full world, with time and space, and thus the accumulating structures, functions, and changing probabilities ("chance"?) that make up the evolutionary story, it has prodigious possibilities. Formalism that obliterates that fecund "substrate" can block the way to more inclusive accounts. We should remember what the formalism invites us to forget. Shifting topic but not principle, I offer a remark attributed to Max (quoted by P. Krugman, New York Times, March 27, 2006) Frisch: "We wanted a labor force, but human beings came".

Acknowledgments I am grateful for exchanges with Elihu Gerson, Peter Godfrey-Smith, William Grassie, John Haught, Mohan Matthen, Nick Matzke of the National Center for Science Education (www.ncseweb.org), Gregory Mengel, Thomas Pradeu, Barbara Herrnstein Smith, and Rasmus Winther.

Chapter 4
Abstractions, Idealizations, and Evolutionary Biology

Peter Godfrey-Smith

4.1 Introduction

Many philosophers agree that *idealization* and *abstraction* are key aspects of science, especially scientific work based on the construction and assessment of models. And while much of the initial philosophical work in this area was concerned with physics, it has become clear that biology, especially evolutionary biology, is another area in which these phenomena play a very significant role. But despite a consensus on the *importance* of idealization and abstraction, there is not much consensus on what these things *are*, and what exactly they achieve. Here I will take the opportunity to offer some general ideas about idealization, abstraction, and the relations between them, emphasizing applications to evolutionary theory.

4.2 Idealization and Abstraction

These two terms have become quite problematic, and surrounded by an unruly literature in which a number of closely related views have been developed. I will proceed initially by abandoning the terms and focusing on some underlying phenomena. Once these are clear, the terms will be re-applied.

In describing something, one may find oneself doing one or both of the following:

1. Leaving things out, while still giving a literally true description.
2. Treating things as having features they clearly do not have.

The former is a matter of ignoring things, ignoring detail. The latter involves an act of imagination; we imagine that something is different from how it actually is.

Initially, I will treat the latter as equivalent to imagining the existence of a fictional thing that is similar to the real-world object we are interested in. There is a general and important role for imagining non-actual objects, properties, and

P. Godfrey-Smith (✉)
Professor of Philosophy, Harvard University, Cambridge, MA, USA
e-mail: pgs@fas.harvard.edu

A. Barberousse et al. (eds.), *Mapping the Future of Biology*, Boston Studies in the Philosophy of Science 266, DOI 10.1007/978-1-4020-9636-5_4,

systems in science. Scientific model-building is in large part the investigation of fictions. Philosophers have often tried to see the models developed in theoretical science as abstract mathematical objects (Suppes 1960), but I suggest that it is often more natural to see them as having a status similar to that of fictions within literature. That is, I see model-builders as often trying to describe and understand systems that are only imaginary, but which would be *concrete if real* (Godfrey-Smith 2006b). When a modeler says that he or she is trying to understand a particular kind of neural network, economy, or ecology, which is too simple to be found in the actual world, I take the modeler at face value. The modeler *is* trying to understand an imaginary neural network, economy, or ecology. He or she will often try to describe and understand it *using* mathematical tools, but that is not the same as saying the modeler is trying to describe an abstract mathematical object.

Very often, these fictional systems are simpler analogues of real-world systems. And in discussing them, a scientist will frequently talk in a way that "aims" the description directly at the real-world system of interest; he or she will present ideas in the form of a literally-false description of the real system, rather than a true description of a different thing, the imagined system.

Going back to my two-way distinction above: ignoring some features in a description of a system is *inevitable* to some extent in *any* description. The question is only how much is left out, and what is retained. The kind of imaginative act seen in the second operation, in contrast, is always optional. One could try one's best never to do it. One could try to always say only what one thinks is literally true.

I now connect these two phenomena to our vexed terminology. An *abstract* description of a system leaves a lot out. But it is not intended to say things that are literally false. An *idealized* description of a system is a description that *fictionalizes in the service of simplification*, in the way described above. The idealized description is often presented verbally *as* a description of a real system, but not a description that is literally true. It is also possible, I said above, to present this as a specification of an imaginary system that resembles the real system. But one thing that the usual way of talking does is make it clear *which* real-world system the fictional one is intended to function as a simplification of. I will return to the small but significant differences between these two ways of thinking of idealized descriptions below. (I should also note that the "more complex system" of interest might itself sometimes be fictional, but the usual case is that in which the other system is real, and I will only deal with this usual case.)

This treatment of idealization and abstraction is not intended to be completely faithful to the way scientists talk; it is intended to clean up a somewhat disorderly set of relations between the two concepts. Several previous philosophical treatments are at least close to this view, and Martin Thomson-Jones (2005) presents a view that is essentially identical, differing only in points of emphasis. Thomson-Jones also helpfully distinguishes this view from related positions in writers like Cartwright (1989) and McMullen (1985). Like me, Thomson-Jones notes that this treatment involves a certain amount of revision of scientific usage.

One thing that leads to confusion here is the fact that both idealization and abstraction can be described as "leaving things out." That phrase can refer to simple *omission* of features, or to an *imagining away* of those features. And crucially, it can be quite difficult to simply omit things, especially in a mathematical context; very often it is necessary to imagine things away. If one wants to "leave out" friction, or genetic drift, and one wants an equation as the form of one's representation, this usually has to involve imagining things absent, even if one offers the verbal comment that one has merely "omitted" these factors.

On the view offered here, what is the relation between idealization and *approximate truth*? The latter is a topic that philosophers have made heavy weather over, especially when seeking to think of a statement's semantic content in terms of its logical implications. Using the more relaxed realist framework I am employing here, it is natural to say that idealized descriptions are in many cases approximately true. Here I understand approximation to truth as a simple matter of similarity between the circumstances depicted and the circumstances that obtain. So when we treat an idealization as a description of some target system, it may, depending on how it is expressed, be approximately true of that system. Further, degree of idealization will correlate inversely with approximate truth. But the relation between the two concepts is not straightforward, because of the evaluative dimension. *Goodness* of idealization is naturally understood in terms of the purposes the idealization serves for us. A description could be very highly idealized, hence scoring low on approximate truth, but be a good idealization in the circumstances, despite this. Conversely, a description could score high on approximation to truth, and be a bad idealization given our goals.

In the light of the ideas above, I can now say something about relations between three closely related scientific activities: offering a description known to be only *approximately true* of its object, offering an *idealized description* of a real object, and building an idealized *model* of that object. First, we saw that idealization involves a departure from reality in the direction of some kind of simplicity – not all approximations and not all models will have this feature. But suppose we concern ourselves only with simplifications. Then I said above that an idealized description of a real object will tend to be approximately true, and also that an idealization can be seen as the specification of a fictional relative of the real object. So once simplification becomes the goal, it seems that approximation, idealization, and modeling are very closely tied together – hard, in fact, to pull apart.

I see the differences between them as in large part practical or strategic. Talk of "approximation" is natural when the description is closely shadowing a real system (at least in intention), and there is little role for the deliberate and rich imaginative construction of non-actual features. Talk of "modeling" is most natural when the scientist's immediate focus is the fictional system itself, relations to the real system are secondary, and the differences between the two are substantial (see also Weisberg 2007). Talk of "idealization" can be natural within *either* of these kinds of activity. So I do not see the mere presence of idealization as an infallible indicator of a model-based style of science, though any idealized description can, *post hoc*, be considered as the specification of a model system that might be investigated in its own right.

To illustrate some of the ideas in this section, I will briefly look at the simplest formal models of evolution, such as one-locus population genetics. It is often said that these models are "abstract," but often said that they are "idealized." It is not really possible for a single description to be both idealized and abstract at once, with respect to the same features of the object being described. So which are they? As I see it, these models do exhibit both phenomena, but with respect to different features.

Suppose we consider a deterministic random-mating one-locus diploid model of natural selection of the simplest kind (Roughgarden 1979), and declare that we treat it *as* a representation of evolution in some real population in which evolutionary change is occurring at one locus. We take the descriptions specifying the model and treat them as descriptions of the empirical system. Then we get a mixture of idealization and abstraction, even assuming that the match with respect to the parameters is as good as it could possibly be for a model of that kind and a real-world system.

The description of the population in terms of gene frequencies at one locus is an abstraction – although a more unusual one than it looks, as I will discuss below. But the model makes an assumption of infinite population size; that is an idealization. The model also assumes random mating; that may perhaps be literally true and an abstraction, depending on the interpretation of probabilities, but it is much more likely to be an idealization. The model assumes no mutation; almost certainly an idealization. What about the fitness parameters? If they are constants in the model, then it is *almost* inevitable that they are idealizations. They are being treated as independent not only of frequency change but also of exogenous environmental change, which will almost always occur to some extent. In many cases, but not in all, the assumption of non-overlapping generations will also be an idealization. So simple evolutionary models are both idealized *and* highly abstract, but in different respects. And strictly speaking, *any* degree of idealization is enough for us to say that the description as a whole is not literally true.

I will make a brief comment on two other technical tools. The "Price equation" is becoming more and more popular as a foundational treatment of evolutionary processes (Price 1972, Frank 1998, Okasha 2006). It is an interesting case. Any evolving population in which there is asexual reproduction at the lowest level (which includes copying of alleles) can be *abstractly* described by a Price equation. The partitioning operation achieved by the equation does not amount to an idealization, and the equation does not use an infinite population size assumption. The fitness parameters are said to hold only for one generation and to represent "realized" fitnesses. All of this can be treated as a literal description. Other theorists, especially in the recent European tradition, like to treat the "replicator dynamics" as a foundational model (Hofbauer and Sigmund 1998, Nowak and Sigmund 2004). Here we are very much in the realm of idealization, especially when assumptions of infinity are made. So some of the alternative foundational models in evolutionary theory at the present time represent slightly different choices with respect to idealization and abstraction.

4.3 Successes and Pitfalls

Abstraction, as described above, is inevitable in all description. So there is no news in the idea that abstraction is essential to scientific work. Without it, communication itself would be impossible. That leaves a significant question, for each field and each question, of how *much* abstraction is appropriate. Idealization I see as near-enough essential to science also, but in a way that is more controversial and substantial. The practice of science shows that fictionalizing is often a crucially important strategy, and also something that people can do while not recognizing what they are doing, and perhaps while talking in misleading ways about what they are doing. But we can certainly imagine a person policing their scientific practice in a way in which all literally false claims are avoided, purged, replaced as soon as possible. The embrace of idealization is a substantial matter, and not something inevitable.

Do these operations generate characteristic kinds of error or wrong turn? Abstraction I said is always present, so the idea of a *characteristic* error is not straightforward. But we might say that there are characteristic errors that can come with extreme or unobvious forms of abstraction. A clear possibility would be a process in which one deliberately leaves some features out, notes the usefulness of the representation that results, and infers that the things abstracted away from do not exist at all. But in this paper I want to emphasize a more subtle kind of error, the reification of the products of abstraction. It can be easy to describe the results of an operation of abstraction in terms that suggest the existence of an additional *entity*. Examples that might be given here will often be tendentious, because people often think this sort of move is justified, not a fallacy. But I would cite the case of propositions, as they figure in much philosophy of language. We note the way that many sentences can have a similarity of meaning, and then come to talk of the "the meaning" of those sentences. Really this is a grouping via abstraction. But the process can lead to a reification of the thing, the meaning, which is then called a proposition. This is seen as an abstract *object* that the sentences all have some special *relation* to. I think this is a wrong turn, and a kind of wrong turn that will figure again in the final section of my paper. As Thomas Pradeu suggested to me, the idea of such a pitfall also has links with Berkeley's criticism, in his *Principles of Human Knowledge* (1710/1965), of John Locke's treament of abstract ideas.

The "reification of abstractions" is sometimes seen as a characteristic error in some theoretical work in biology – Lewontin's critiques of ANOVA and heritability methods (1974) are examples. How does this fit into the framework used here? The situation alleged, in such a critique, is something like this. An abstract description yields a summary, such as a statistic. The statistic is then taken to pick out something like a *discrete causal factor*, a causal player of its own, something that is part of the machinery. The *grammar* of description appropriate to a discrete causal factor is applied. In the most problematic cases, it is treated as an entirely different kind of entity, with special relationships to the other, ordinary causal players.

Idealization certainly generates a characteristic error close to the first one discussed in the case of abstraction. We pretend that some features of a real system do not exist, for the purposes of simple and compact representation. Finding ourselves doing well with the resulting compact description, we infer that that there is nothing more to the system that what has been recognized in our idealized treatment. I have argued that this is a characteristic of some philosophical work in metaphysics (2006a). Here I follow much older work by John Dewey, who described a similar problem in different terms, in Chapter 1 of *Experience and Nature* (1929/1958).

But to conclude this section, it is worth emphasizing how well things can go for a scientific discussion, when it features the right mix of highly idealized work and more concrete particularistic work. In evolutionary theory and related parts of biology, this sort of interaction has a long history. Darwin's *Origin of Species* (1859) is notable for the very slight role for idealization. His focus is squarely on the concrete nature of organisms, and the actual processes in which these organisms are involved. He does offer abstract claims, and entertains abstract hypothetical propositions from time to time. Darwin was also willing to formulate some principles in the form of "laws" (see the "Recapitulation" of the *Origin*). But in part because he does not try to formalize these principles mathematically, he does not find himself idealizing away from the actual nature of organisms very much, and the work is also much less abstract than much of the work that came later. It was partly this focus on the empirical and concrete that made the *Origin* so powerful and convincing when it appeared. But idealized work, including mathematical modeling, appeared quickly. Francis Galton's "Typical Laws of Heredity" (1877), for example, looks very much like a piece of model-based science. It was followed by similar work by himself and others within the "biometric" school (Provine 1971).

Many of the landmark works of the 1930s, such as R. A. Fisher's and Sewall Wright's work, have an entirely different character from that of the *Origin*. Many of Fisher's crucial contributions were made via his ability to pick the right simplified fictional system to think about, in order to fully understand how a process works in the more complex empirical domain. He thinks about populations whose size is effectively infinite, that are unstructured in space (or not affected by any structuring that may exist), and whose members differ with respect to the simplest kinds of genetic features – usually one-locus genotypes.

People sometimes side with one style of science or the other. Fisher's approach might be lampooned as "bean-bag genetics" and seen as an exercise in mere mathematical fantasy. It is less common to disapprove of Darwin's style of biology, but one does hear occasional murmurs of this kind. Darwin's work is, to the formal mind, a great unruly mess of empirical detail. However, I intend this pair of cases to be an illustration of how things look when things go well. The two styles of evolutionary theorizing seen here are surely complementary; each did things that the other could not have done, and our understanding of evolutionary processes draws essentially on both.

I think we see some of the same kind of complementarity in some recent work on the "major transitions" in evolution. Leo Buss' *The Evolution of Individuality* (1987) is a Darwin-like work. Entirely non-mathematical and crammed with empirical detail, it spurred much of the later interest in the problem. Maynard Smith and Szathmáry's *The Major Transitions in Evolution* (1995) is a model-based book full of idealizations. So we have two books written around the same time, on the same topic, and of similar importance, where one book applies a modeling strategy and the other does not. This gives us a nice illustration of the contrasts between the *style* of science that uses models and the style (or one style) that does not. The Maynard Smith and Szathmáry book is also noteable because although it is a model-oriented book, most of the models are *not* presented mathematically. The exposition proceeds through a constantly shifting mixture of words, pictures, and a few equations here and there. So this book is a useful corrective to the view that model-based science has some essential connection with mathematical techniques.

Maynard Smith and Szathmáry's book can be contrasted in this respect with Rick Michod's book *Darwinian Dynamics* (1999). Michod is concerned with similar topics again. (He is close to Buss in his focus.) Michod's book is model-based, but in this case the models *are* given in explicit mathematical form. So recent work on the major evolutionary transitions shows what I see as a complimentary mix of model-based and non-model-based approaches, and the modeling work itself includes work that is primarily mathematical and work that is not.

4.4 The Informational Gene

In the rest of this paper I will use the framework outlined earlier to comment on what might seem to be an unlikely case, the informational treatment of genes in biology. I take it that this is a striking and philosophically interesting feature of recent genetics and evolutionary theory. Different aspects of the phenomenon can be distinguished. One is the claim that genes code for, or contain information specifying, whole-organism phenotypes or the course taken by developmental processes. Here, we see an unusual and philosophically controversial description of the causal role of genes, but the genes themselves are taken to be ordinary material objects. More recently, it has become quite common to insist that we should, further, see genes themselves as in some sense *made* of information, as informational objects, at least in the context of evolutionary work. So the informational mode of description is now not just being used in the specification of causal roles, but also in the specification of what kinds of things, what kinds of objects, figure in evolutionary theory. The most dramatic expression of this kind of thinking is seen in George Williams' 1992a book *Natural Selection*. There he claims that we have learned, via evolutionary theory, that information is one of the basic ingredients of the universe, along with matter and energy, inhabiting its own "domain."

In earlier work I have criticized the less metaphysical kind of informational description of genes – the use of semantic concepts in the specification of causal roles (Godfrey-Smith 2000b). My argument has been that there are good reasons to take genes as coding for, or carrying information specifying, the primary structure of protein molecules, but there is no reason to use this semantic language to describe the role of genes in the causation of any product downstream of those protein molecules (see also Griffiths 2001). In this discussion I will comment on the more tendentious and, to my mind, extravagant version of the informational treatment of genes, the idea that there is such a *thing* as the informational gene, as well as the concrete DNA molecule, or that genes for evolutionary purposes informational *entities*. I think this is misconceived idea, and one that can be understood as the misapplication of a process of abstraction.

Let us first ask how the idea of an informational treatment of genes as objects arises in evolutionary theory. It arises in order to achieve a kind of *grouping* operation. Considering an actual-world population, we find it is composed of a collection of particular organisms containing a great mass of particular molecules. Theory has taught us that we can get substantial purchase on change in this population by attending to a particular kind of molecule in the cell nucleus (DNA). And our "attending" to it involves a peculiar kind of focus. We attend to individual chunks of this molecule that are not physically separated from other chunks and their surrounds. We group token chunks of this molecule into types according to base sequence, and then characterize the whole population by the frequencies of the types. Thus we get the evolutionarily important measure of a gene frequency at one locus. We then pay close attention to changes in these frequencies, by attending to the rates at which new molecules of a given kind are produced via template processes from old ones.

All this involves a more peculiar counting procedure than is often supposed. In a population with a germ line, we ignore almost all tokens of the molecules in our counting. We do the grouping of tokens into types by attending to the sequence of bases, but we are allowed to ignore some base differences ("silent" differences, for instance). The demarcation of "chunks" and other aspects of the counting is also dependent on the peculiar process of "crossing-over," occurring at a particular stage in the cell cycle.

What we have done is engage in a highly abstract description of both the organisms and the population. At the heart of this abstract description, again, is a *grouping* operation, one that groups token molecular structures separated by both space and time. A great many token molecular structures are treated as falling into a common type, in virtue of their sequence properties. There is nothing wrong with this, but it sets us up for a possible additional move that is mistaken. Rather than saying that all there is the collection of molecules themselves, which we describe abstractly for certain purposes, we might be tempted to say there is an extra entity, an abstract additional thing, that the particular molecules have some important relation to. That is, we could be tempted to say that when we talk of genes for evolutionary purposes, we mean to describe the doings of an abstract entity, the "informational gene." For me, this is just like the unnecessary and misleading move that many philosophers writing about language make, when they reify "propositions." They note the phenomenon of similarity or identity of meaning across sentences (type and token), and conceive of this in terms of relations between the sentences and an extra entity, the proposition, that is real but abstract.

In the biological case, the shift between the operation of grouping material objects for purposes of abstraction, and the introduction of an extra entity that bears some relation to the material things, is illustrated in this quote from David Haig (1997).

> *Gene* ... can refer to the group of atoms that is organized into a particular DNA sequence — each time the double helix replicates, the gene is replaced by two new genes — or it can refer to the abstract sequence that remains the same gene no matter how many times the sequence is replicated. The *material gene* (first sense) can be considered to be a vehicle of the *informational gene* (second sense).

I don't think that Haig means to put anything like the same metaphysical weight on the informational gene that people like George Williams do. In practice, Haig regards this operation of grouping, and all the concepts used here, in very pragmatic ways. That is clear from his paper. But the road to an error of reification is shown very clearly in the language here. We seek to group a number of tokens, in virtue of a similarity of sequence: fine. But that does not require anything like the introduction of an extra entity for which the token material objects function as "vehicles."

Perhaps the point might be put by saying that in the case of *other* molecules, like sugars and lipids, we are perfectly able to group them into types without introducing an extra abstract thing that the material molecules are "vehicles of." There is no "informational lipid," grouping all the lipids of a given structure; there is just a collection of molecules that we have reason to group as a kind. Genes do play a different causal role from that of lipids, because of the importance of exact DNA sequence in protein synthesis, and the importance of DNA in inheritance. I know that this is supposed to mark a key difference between DNA and things like lipids. But those differences in causal role are not of a kind that motivate an entirely different ontological treatment of genes. They are just differences in the concrete causal processes in which the two classes of molecules figure.

Why does this matter? First, if one cares about ontology then it seems to simply be an erroneous move to treat genes in a de-materialized fashion. But there are further consequences as well. Once genes are de-materialized in this way, they are removed from the ordinary causal give-and-take, and can come to be seen as explanatory factors of a special, ultimate kind, telling us our true natures. But all that is real here is the DNA molecules, with their various properties. They are, indeed, extremely important, but important as material things that are part of the causal give-and-take.

Acknowledgments I am grateful to Anouk Barberousse, Marie-Claude Lorne, and Thomas Pradeu for helpful comments on an earlier draft.

Chapter 5
The Adequacy of Model Systems for Evo-Devo: Modeling the Formation of Organisms/ Modeling the Formation of Society

Scott F. Gilbert

5.1 Introduction

The writer Jorge Luis Borges tells us of a wondrous Chinese encyclopedia whose classification of animals begins with the dichotomous branch: Those animals belonging to the emperor, and those animals that don't (see Foucault 1970). In the social scheme of scientific organisms, a similar wisdom prevails. There are those privileged imperial organisms called model systems, and then there are all the other organisms. There are seven basic model systems of developmental biology: the fruit fly *Drosophila melanogaster*; the nematode *Caenorhabditis elegans*, the mouse *Mus musculis*, the frog *Xenopus laevis*, the zebrafish *Danio rerio*, the chick *Gallus gallus*, and the mustard*Arabidopsis thaliana*. For many researchers, having their experimental organism be considered a model system is an important goal. The recognition that one's organism is a model system provides a platform upon which one can apply for funds, and it assures one of a community of like-minded researchers who have identified problems that the community thinks are important. There has been much lobbying for the status of a model system and the fear is that if your organism is not a recognized model, you will be relegated to the backwaters of research. Thus, "model organisms" have become the center for both scientific and political discussions in contemporary developmental biology.

In 1999, I received the following plea from an eminent chicken embryologist:

> There apparently is going to be a meeting at the NIH to discuss genomics of non-traditional model organisms. It seems they are almost completely ignoring the chick system... they view it as an agricultural, not experimental animal... If you can spare the time, please write a letter of support (no matter how brief) to Dr. Paul Goetinck as soon as possible supporting this case for funding a chick genome project from NIH. It is a unique opportunity to establish the chick as one of the acknowledged 'model systems' that NIH will fund, and should also improve the funding for non-genome related chick research grants.

The fight to establish chicks as an official NIH model system continues, as new techniques (such as electroporation of dominant negative genes and the creation of

S.F. Gilbert (✉)
Howard A. Schneiderman Professor of Biology, Swarthmore College, Swarthmore, PA, USA
e-mail: sgilber1@swarthmore.edu

A. Barberousse et al. (eds.), *Mapping the Future of Biology*, Boston Studies in the Philosophy of Science 266, DOI 10.1007/978-1-4020-9636-5_5,

chick embryonic stem cells) not only circumvent the original genetic difficulties in studying chick development but might synergistically interact with the traditional advantages of the chick for grafting and lineage tracing (Stern 2005). Being declared a "model system" is a political as well as scientific endeavor.

As would be expected when there are both financial and scientific stakes involved, there will be controversies over which animals provide the best model for a particular phenomenon. For instance, the origin of mesoderm and bilateral symmetry has long been a fight between the people supporting the acoel flatworms (barely mesodermal and definitely bilateral) with the people supporting hydra (not yet either mesodermal or bilateral). Does one look at the organism on the positive side of the transition (Cook et al. 2004) or the negative (Meinhardt 2002; Fujisawa 2006)? As a model system to study the origin of vertebrates, there are two major contenders among the non-vertebrate chordates. There are the advocates of the cephalochordate amphioxus being the sister clade to the chordates (Holland 2002; Holland et al. 2004), and there are other investigators who are putting their money and livelihoods on the urochordate ascidians (Corbo et al. 2001; Wada 2001; Passamaneck and Di Gregorio 2005; Delsuc et al. 2006).

These debates represent attempts by groups of scientists to position their particular organism as the species most suitable to answer a particular set of important scientific questions. The positioning of any particular organism is based on scientific principles but is also crucial to the funding and visibility of these laboratories. Indeed, the positioning of certain*questions* has become a political issue, and it is intimately linked to the model systems that may make certain questions more answerable than others. Thus, the choice of model systems has been very important in the history of biology. The study of regeneration (and its inclusion in the normative of development and genetics) underwent a precipitous decline after T. H. Morgan campaigned against its usefulness as a model system to study heredity, casting it in opposition to Drosophila (see Mitmann and Fausto-Sterling, 1992).

The National Institutes of Heath in the United States defines model organisms as follows (http://www.ncbi.nlm.nih.gov/About/model/mammal.html):

> Over the last century, research on a small number of organisms has played a pivotal role in advancing our understanding of numerous biological processes. This is because many aspects of biology are similar in most or all organisms, but it is frequently much easier to study a particular aspect in one organism than in others. These much-studied organisms are commonly referred to as model organisms, because each has one or more characteristics that make it suitable for laboratory study. The most popular model organisms have strong advantages for experimental research, such as rapid development with short life cycles, small adult size, ready availability, and tractability, and become even more useful when many other scientists work on them. A large amount of information can then be derived from these organisms, providing valuable data for the analysis of normal human development; gene regulation, genetic diseases, and evolutionary processes.

At the same time as the NIH is seeking to fund model organisms, the more basic science-oriented National Science Foundation has initiated two programs that*critique* the model systems approach and explicitly seek to fund investigations into non-traditionally used organisms. At stake are the notions of what constitutes normal development, how macroevolution occurs, and how much of what we know about development is an artifact of our choosing systems that are each quite similar. The canonical organisms for developmental biology have all been selected for their small size, large and early litters, immediate separation of the germline, and little or no environmental regulation of development. This, Bolker and others (Buss 1987; Bolker 1995; Bolker and Raff 1997; van der Weele 1999; Gilbert 2001; Wilkins 2002; Poccia 2006) have claimed, distorts what nature really is about. Most organisms won't grow in the laboratory, take a long time to mature, and are susceptible to all sorts of environmental influences. What is good for the NIH seems to be bad for the NSF. Thus, the "model systems approach" is presently at the heart of numerous controversies in developmental biology.

The reason for the small number of developmental model system has to do with the alliance between developmental biology and genetics. When developmental biology had been a physiological science, it had different models—newts, sea urchins, Ranid frogs, ambystomid salamanders, slime molds, flatworms, and chicks. But when it became a genetic science, the environment had to play a smaller role. To get a good genetic story, one does not want to deal with organisms whose phenotype is significantly controlled by the environment. As Sonia Sultan (2003) points out, "neo-Darwinian botanists were often quite frustrated in their attempts to discern genetically based local adaptations through this 'environmental noise,' " and this led them to overlook the adaptive nature of these developmental plasticity. In the zoological side of developmental biology, the desire to link developmental biology with genetics (and the desires to easily breed the animals) led to the adoption of a very limited number of model species, each of whom had been selected for the lack of significant environmental components to the phenotype (Bolker 1995; see NIH statement above). It is interesting to read the preface to volume 1 of *Current Topics in Developmental Biology*. It was written, not by an embryologist, but by a geneticist, Joshua Lederberg (1966). He proposed that if developmental biology were going to make progress, it would have to have a model organism like *E. coli* to work on. He suggested two such models: the mouse (as a surrogate for the human) and "some very simple system like a rotifer or nematode".

Evolutionary developmental biology and its sub-discipline, ecological evolutionary biology, are both claiming that the canonical model systems are good starting points for evolutionary and developmental investigations, but that they give a biased view of nature. First, these animals can give one the erroneous impression that everything needed to form the embryo is within the fertilized egg. Second, these animals in the laboratory do not provide adequate explanations for the way animals develop in the wild. Tadpoles in the wild look different from tadpoles reared in aquaria, and their phenotype develops, in part, from cues emanating from competitors and predators. Environmental chemicals that are harmless in the laboratory may be dangerous to developing organisms in the wild (Colburn et al. 1996; Hayes et al. 2003; Relyea and Mills 2003). Third, as Neff and Rine (2006) note, "model organisms have become a 'comfort zone' for biologists, luring them away from investigating questions that cannot be answered with any of the existing models." And fourth, the organisms used for modeling a particular phenomenon might be idiosyncratic. Species can be defined as those organisms that develop in a particular way, using particular molecules and processes. Thus, the development of no single organism, by this definition, can circumscribe the developmental attributes of its clade. Most arthropods probably do not use a gradient of Bicoid to form their head, even though this gradient is remarkably important for *Drosophila*. Most amphibians probably do not form their mesoderm as *Xenopus* does, even though *Xenopus* has become a model for mesoderm formation. The mouse is a good starting point for studying other mammals; but mammalian development has diverged enormously and certainly beyond what one murine species can define (see Benirschke 2006). David and Marylyn Kirk (2004) have spent their professional lives sorting out the ways that Volvox distinguishes germ line form soma and have come to the conclusion that "*Volvox carteri* is an excellent model for other *Volvox carteris*." Even other Volvox species do this important act in different ways.

5.2 New Model Systems for Evo-Devo

With the advent of evolutionary developmental biology (evo-devo), new questions are being asked, and they require new model systems (see Wilkins 2002). This is an interesting phenomenon, since evo-devo has been characterized by the revolt against model systems. The existing model systems have been found to be inadequate; so *non-model* systems have been utilized to address these problems. (In my research, I am studying turtle embryos. These have all the disadvantages that model systems try to circumvent. They have few mutations, have a long gestation time, a long duration until sexual maturity, few embryos per litter, and they are available only three months of the year.) Now, however, as evolutionary biology becomes stabilized around certain questions, new model systems are emerging.

Three major questions in evolutionary developmental biology concern the following issues:

(1) What are the genes whose small changes can make large phenotypic differences?
(2) How did the basic properties of most organisms—three germ layers and bilateral symmetry—arise?
(3) How does the environment control the genes that produce phenotypic diversity within developmentally plastic species, and what accounts for this plasticity?

Each of these areas is attracting new investigators and new organisms. The dog *Canis familiaris* and the three-spined stickleback fish are being considered as model systems for studying altered gene expression during evolution. Moreover, traditional model systems such as zebrafish and *Drosophila* are being "repositioned" to become aids to the discovery of specific differences within lineages. Thus, zebrafish is being seen as a useful source of genes through which the evolution of specific piscine lineages might be studied (Schilling and Webb 2007). The cnidarian *Nematostella* is being seen as a model system for looking at the origins of the bilateria. And, last, the dung beetle *Onthophagus* is being proposed as a model system for studying the evolution and the properties of developmental plasticity.

5.2.1 Dog Development as the Model System for Evolution by Altered Gene Expression

Despite what Michael Ruse (this volume) might think, Darwin (1883, p. 282) was well aware that natural selection could not account for the origin of variation. Other evolutionary biologists (starting with Thomas Huxley; see Gilbert 2003c, 2006a) have noted that evolutionary novelty arises from changes taking place during development. Natural selection can delete or preserve such new phenotypes. Darwin (1859) also recognized that artificial selection is a model of evolution wherein harsh selective pressures imposed by human selection and mating regimes can rapidly change the appearance of an organism. One of the best examples is that of dogs: There are now over 200 recognized breeds (and about 1000 local breeds) of dogs, each derived from *Canis lupus*, the wolf, starting about 15,000 years ago (Savolainen et al. 2002).

Dog breeding, write Neff and Rine, "has been an ongoing experiment in the rapid evolution of form and function." Moreover, the completion of the canine genome has made *Canis familiaris* "genetically tractable and poised to offer insights into evolution, development, and behavior." These authors point out that while null mutations such as those readily produced in the mouse can tell you about how a system breaks down, such mutations are not usually relevant for understanding natural diversity or evolution. In dogs, however, you have remarkable diversity of functioning systems. Dogs can differ 50-fold in mass and have behaviors ranging from completely docile to overtly vicious. Moreover, these differences are inheritable. There are not only breed-specific temperaments, but even dog breeds that perform the same behaviors (such as herding) do it differently from one another. The variation that dogs have is very different from the variation produced in the laboratory on caged mice. To study this normal and enormous variation is critical if one wishes to study evolution or, for that matter, brain function.

The central argument for the dog as model system for evo-devo is that it has enormous variation, that these variants are functional, not pathological, and that this variation is occurring within the same species. Stockard (1941) and more recently others (Gilbert 1991) have pointed out that the difference in dog snouts represent remarkable changes in the migration and proliferation rates of cranial neural crest cells. Now that the canine genome has been completed (from a female boxer; Lindblad-Toh et al. 2005), it is hoped that comparisons can be made. There are over 50 million pedigreed dogs in the United States, and there should be plenty of molecular variation to map. The goal is to elucidate the genetic variations that underlie the different morphologies and behaviors that define each breed.

And such studies have already been started. Fondon and Garner (2004; Caburet et al. 2005) have shown that length-variation of tandem repeats in protein-coding regions of developmental genes are associated with morphological changes in dog breeds, and the gene homologous to human *TCOF* (whose mutations cause the Treacher-Collins syndrome of facial shape anomalies) shows variants highly associated with head depth in dogs (Haworth et al. 2001).

The dog also has the advantage of being our companion species. Haraway (2003) claims that the dog and human form a coevolving unit, and this is echoed by two researchers from the National Human Genome Research Institute of the NIH. Sutter and Ostrander (2004) conclude that, "More than simply our servants, they are a species that has evolved in harmony with us, so that encoded within their genome is the ability to read both our hearts and minds, humbling us, but never diminishing us." Thus, the dog is poised to be a model for the portion of evo-devo which seeks to find in the changes of developmental regulatory genes the mechanisms of large anatomical change during evolution. Moreover, the dog genome is seen by some as being a "companion" genome that can offer clue to human health and even to the human condition.

5.2.2 Nematostella: At the Origin of the Bilaterians

There are several model systems being described as representatives of basal phyla. Although all phyla are derived (see Jenner 2006), some appear to have retained characteristics that may be ancestral to both the deuterostome and protostome lineages (see Tessmar-Raible and Arendt 2003). One of the "new" model systems to study the origin of more sophisticated lineages is the cnidarian. *Nematostella vectensis*, the starlet sea anemone. Its proponents claim that it represents two of the most fundamental transitions in animal evolution: the origin of bilateral symmetry and the origin of the mesoderm. A *Nematostella* website (http://www.nematostella.org/) claims: "The starlet sea anemone, *Nematostella vectensis*, is becoming an increasingly important model system for the study of development, evolution, genomics, reproductive biology, and ecology." When Martindale et al. and colleagues (2004) introduced *Nematostella* as a model system for the study of triploblasty, they proposed it for the reasons traditionally used to justify such a designation: simplicity, ability to be cultured, large number of embryos, availability of embryos all year, and rapid development of the embryo:

> *Nematostella* has many practical advantages as a developmental model, including a simple body plan and a simple life history. It is a hardy species, easy to culture (Hand and Uhlinger, 1992) and will spawn readily throughout the year under laboratory conditions (Fritzenwanker and Technau, 2002; Hand and Uhlinger, 1992). Sexes are separate and fertilized embryos develop rapidly to juvenile adults bearing four tentacles.

To demonstrate the usefulness of *Nematostella* as a model organism for looking at the origin of triploblasty and bilateral symmetry, Martindale and his colleagues have shown that these organisms possess the typical bilaterial body plan (common to vertebrates and insects), but in a rudimentary form. Thus, the genes for dorsal-ventral polarity (BMP and chordin) are found, but they appear to be playing slightly different roles than in the more highly specialized bilaterians (Matus et al. 2006); the genes used in insects and vertebrates for germ-cell specification are found there, too; but seem to be playing more roles than expected (Extavour et al. 2005). The finding in *Nematostella* of many of the transcription factor families known to be critical in the development of contemporary insects and vertebrates has given further reasons to look at this organism as an example of an organism that is ancestral to all the major lineages of the animal domain (Magie et al. 2005):

> The diversity of expression patterns among members of these gene families in *Nematostella* reinforces the notion that despite their relatively simple morphology, cnidarians possess much of the molecular complexity observed in bilaterian taxa.

In 2004, Finnerty and colleagues made the following announcement:

> We show that *Nematostella* uses homologous genes to achieve bilateral symmetry: Multiple Hox genes are expressed in a staggered fashion along its primary body axis, and the transforming growth factor-beta gene *decapentaplegic* (*dpp*) is expressed in an asymmetric fashion about its secondary body axis. These data suggest that bilateral symmetry arose before the evolutionary split of Cnidaria and Bilateria.

Thus, bilateral symmetry can first be seen in the Cnidarians. Moreover, so can mesoderm. Not only are there muscular cells among the cnidarians, but these cells are expressing the "mesodermal genes" that characterize mesodermal specification in insects and vertebrates (Martindale et al. 2004).

Martindale (2005) writes that the field needs these models to place renewed emphasis on the functional interactions of complex gene-regulatory pathways in a phylogenetic context so that scientists could "unravel the legacy of morphological complexity that is seen in the animals of today." This is echoed by Darling and colleagues (2005), who summarize the importance of *Nematostella* over other basal organism models:

> In recent years, a handful of model systems from the basal metazoan phylum Cnidaria have emerged to challenge long-held views on the evolution of animal complexity. The most-recent, and in many ways most-promising addition to this group is the starlet sea anemone, *Nematostella vectensis*. The remarkable amenability of this species to laboratory manipulation has already made it a productive system for exploring cnidarian development, and a proliferation of molecular and genomic tools, including the currently ongoing *Nematostella* genome project, further enhances the promise of this species. In addition, the facility with which *Nematostella* populations can be investigated within their natural ecological context suggests that this model may be profitably expanded to address important questions in molecular and evolutionary ecology. In this review, we explore the traits that make *Nematostella* exceptionally attractive as a model organism, summarize recent research demonstrating the utility of *Nematostella* in several different contexts, and highlight a number of developments likely to further increase that utility in the near future.

Even attractive models are nothing without their agents.

5.2.3 The Dung Beetle Model System

One of the major changes in developmental biology over the past decade has been the recognition that the environment plays an instructive role in phenotype production. Polyphenisms, norms of reaction, and developmental symbioses, long a part of ecology, are now becoming seen as being normative for developmental biology. What had been a province of exceptions is becoming the rule, as mammalian gut development has been found to be symbiotically regulated, and as evolutionarily cued epigenetic methylation is seen to alter DNA in numerous animals, including us mammals (see Gilbert 2004).

This has been reflected in a workshop document published by the National Science Foundation of the USA (NSF 2005). This booklet specifically relates the influence of model systems in directing the course of developmental biology.

> Developmental biologists have, for many years, focused their efforts to understand ontogeny by selecting a few model organisms that are genetically tractable, and that are appropriate to the study of fundamental processes of development at the genetic, molecular and cellular levels. These efforts have led to a detailed understanding of the genetic mechanisms that are involved in the control of developmental events. Many of the findings that have emerged from this work have proven remarkably transferable among the models studied. Developmental biologists have relied on model systems with relatively little but controllable genetic variation. Consequently they have typically not studied the way developmental mechanisms differ among species, nor the variance in mechanism among individuals due to normal variation in genetic and environmental factors. Some developmental biologists have recently begun to expand their studies to include non-model species for understanding aspects of developmental processes not reflected in the models. Still others are interested in illuminating the breadth or limitations of the generalizations discovered in the model systems. Recent developments in genomic approaches have facilitated this move away from the few genetically tractable model systems.

The NSF authors contrast this to the physiological approach:

> Animal physiologists, by contrast, have been reluctant to adopt the use of a relatively small number of model species. This is in part because the physiological principles that bind the subscience cohesively, such as regulation and control of the functions required for normal operation, are known to differ between species.

The NSF then calls for an Integrative Developmental Biology (IDB) that would synthesize the methodologies, analytic tools, and conceptual approaches of these two disciplines.

So how does one study ecological developmental biology ("eco-devo") and the effects of environmental agency on normative development? First, one needs an animal with a readily identifiable suite of traits that change consistently with the environment. Two claimants for the eco-devo model system have recently come forward. The first contestant, from Paul Brakefield's laboratory, has been the Malawian butterfly *Bicyclus anyana* (see Beldade et al. 2002, 2005). In this butterfly, temperature helps determine phenotype. During the cool dry season, the butterfly walks among the leaf litter, and its cryptic brown coloration protects it. During the hot, moist months, the butterfly flies, and its eye-spots protect it from insect predators (Lyytinen et al. 2004). *Bicyclus* thus provides an excellent system for looking at phenotypic plasticity.

Moreover, by combining forces with Sean Carroll's developmental genetics laboratory, the molecular mechanisms for this plasticity are being found (Brakefield et al. 1996). The temperature causes an increase in the ecdysone hormone during a particular stage of larval development, and this hormone is able to sustain the expression of the *Distalless* gene in the presumptive eyespots of the wing imaginal disc. The Distalless protein activates a series of transcription factors that initiate color development throughout the wingspot in a concentric manner. The ability to transform the butterfly by molecular means, study its physiology, monitor its development, and analyze its ecology and evolutionary biology has made this a particularly exciting species to follow. As Beldade and colleagues (2005) remark, "This system provides the potential for a fully integrated study of the evolutionary and developmental processes underlying diversity in morphology."

While *Bicyclus* is being used by several investigators, other model systems are also being proposed to study developmental plasticity. One of the most fascinating examples of developmental plasticity involves the structural and behavioral polyphenism in the dung beetle, *Onthophagus*. Male dung beetles can be separated into two distinct classes: The large males have head horns, while small males have rudimentary horns or are hornless. Horn length varies allometrically with body length, and this results in a bimodal distribution of horn sizes. Up to a particular body size, the males are essentially hornless. Then, after reaching this threshold, the horn grows much faster than the body. Body size is determined entirely by the size and quality of the dung provided the larva by its mother. When a larva runs out of food, it metamorphoses into an adult (Emlen 1994, 1996). The regulation of horn size by food is regulated through the prepupal endocrine system, wherein ecdysone synthesis stimulates horn growth (Hunt and Simmons 1998; Emlen and Nijhout 1999).

The hornless and horned males have very different sexual strategies (Emlen 1997; Nijhout 2003). The horned males defend tunnels that are dug by the females and use their horns to fight other males who want access to these females. The males with the longer horn win. The hornless males would seem to be at a reproductive disadvantage; but not only are the horns polyphonic, so is a behavior. Instead of fighting (in which case they would certainly lose), the hornless males either try to sneak past a defending male or actually dig their own tunnels into the tunnels of the females. This polyphenism results in divergent selection: large males benefit from large horns (since it helps them win combats), while short males benefit from the smallest possible horns (because horns get in the way of digging and sneaking).

Thus, the polyphenism in dung beetles involves the coordination of both morphological structures and behavioral strategies by the endocrine system. But if you're a male dung beetle, your body and attitude are due largely to the amount of dung your mother left you. Emlen (2000) sees a reciprocity between how the study of beetle development might contribute to evo-devo and how evo-devo might contribute to the study of beetle development.

> In principle, understanding how development affects the expression of morphological traits should explain the evolution of those traits. However, empirical studies demonstrating an immediate relevance of development for understanding evolution in natural populations are rare because most population biologists do not study the developmental mechanisms regulating the expression of their traits of interest. One trait in which to examine this question is in the horns of beetles. The behavior associated with horns, the evolution of horns, and the development of horns have been explored for the same two species; consequently, it is now possible to integrate the results from these studies and to explore how knowledge regarding the mechanism of horn development influences our understanding of beetle horn evolution.

Thus, the dung beetle might be a model system for looking at the evolution of developmental plasticity (of both form and behavior), the consequences of developmental plasticity, and the hormonal mediation by which such plasticity is regulated by environmental factors (Moczeck et al. 2006).

5.3 Model Systems and the Assumptions of Developmental and Evolutionary Biology

Evolutionary developmental biology, especially its ecological component, has ontological, epistemological, and methodological implications for biology and philosophy. First, ecological developmental biology is an important critique of reductionist ontology. Once one gets away from the genocentrically selected model systems, one finds that in most species, the phenotype depends to a significant degree on the environment, and that this is a necessary condition for integrating the developing organism into its particular habitat.

A second critique from ecological developmental biology concerns what counts as an "individual." Ecological developmental biology finds the individual to be more of an ecosystem and that phenotype is often derived from more than one genome (Gilbert and Epel 2009). Our "self" becomes a permeable self. Kauffmann (1995) claimed that "All evolution is co-evolution". We may have to conclude that "All development is co-development."

The notion of developmental plasticity also provides a critique for our ontology of evolution. Evolution is real, to be sure; but the metaphors we use to describe it may need revision. Darwin used the metaphor of metallic wedges. Only so many cuts fit, and when one fits better, it forced the others out. "Fit" is the key metaphor here, and it is similar to the lock-and-key model of enzyme activity that was being proposed at a similar time. However, the lock-and-key model doesn't work. (It doesn't have a good "fit" to nature.) Rather, the current model is that of "induced fit", where there is an interaction between the substrate and the catalytic site. Developmental plasticity indicates that there are interactions between the animal and its environment such that the phenotype will become fit for the environment. It is a kind of "induced fitness."

This idea is the converse of niche construction, and is another example of where the niche and the organism interact (see Odling-Smee, this volume). Indeed, niche construction coupled with developmental plasticity may be the ecological analogue of one of the major principles of embryology, reciprocal induction (Laland et al. 2008). This should not be a surprise; the tangled web consists largely of developmental interactions. The pollination of a flower contains numerous developmental components, such that the flower's development and opening are timed with the life cycle of its pollinators. In some species of gallfly, the larva hatches from its egg and begins to eat the host plant stem. Salivary proteins from the larva induce cell proliferation in the plant stem to form the callus. The gall fly larva resides for the summer in that callus, and when the temperature declines, the gall begins to dessicate. The dessicated gall produces volatile compounds that induce gene expression in the larvae, causing it to synthesize those antifreeze compounds (glycerol, trehalose, sorbitol) that will allow it to get through the freezing winter (Danks 1987; Irwin et al. 2001). Mutual inductions are seen here. The gall fly creates its niche by inducing cell proliferation in the plant. The niche serves not only as a source of nutrition but also protection through its inducing gene expression in the larva. Interestingly, the same might be said about the mammalian embryo, whose trophoblast secretes factors that help create the uterine niche (by evoking the developmental plasticity of the mother), while the mother helps the embryo to develop by providing nutrition, oxygen, and defense.[1]

Some of the obvious developmental interactions involving niche construction come from symbiotic systems wherein the partners mutually induce changes in the other partner's development (McFall-Ngai 2002). The bacteria in our guts obvious have a niche in our intestines, and they have helped create this niche. Mice grown in germ-free environments have poorly developed enteric circulatory systems. The bacteria actually induce gene expression in the mammalian intestinal epithelial cells, causing these cells to make several proteins essential for proper capillary development, immune system development, proper lipid absorption, and proper bacterial containment (Hooper et al. 2001; Stappenbeck et al. 2002; Rhee et al. 2004). They even induce the host intestine (their niche) to mount immune responses against other, more dangerous, bacteria and fungi (Hooper et al. 2003). Developmental plasticity is essential for niche construction and for the organism to respond to changing environments.

Model systems both reflect the questions asked by a discipline and channel the research that is being done. This is the positive feedback loop that makes avoiding model organisms so difficult. Are model systems adequate? I think that we are finding out that this depends on what one constitutes as the "system." A system is that set of parameters which includes all the relevant information needed to explain a phenomenon. If the "system" just refers to the organism and those elements below it (as in Ebert's and Sussex' excellent text, *Interacting Systems in Development)*, then, the answer is no. However, if like Paul Weiss, one can include the environment as part of the system, then the answer might be yes. This approach harkens back to Johannsen's original description of phenotype as "genotype plus environment". The developmental program is dispersed. Although the organism is the central component, whose instructions arose from the DNA that made it through the bottleneck of meiosis and fertilization, the information may be disseminated

[1] While this essay will confine itself to niche formation in the environment, there are numerous cases, as mentioned by John Odling-Smee (this volume), where the embryo makes niches. Probably the most well studied is the "stem cell niche." Here, a particular environment is created by intercellular interactions such that cells residing in this niche retain their ability to regenerate themselves as well as producing a more differentiated cell population (Spradling et al. 2001). That selection occurs within the embryo was mentioned by Wilhelm Roux (1881) (who probably did see it as Darwinian natural selection) and became a fundamental influence on Nietzche's view of the body politic (Müller-Lauter 1978; Moore 2002). A more sophisticated notion of selection occurring within embryos was given by Waddington (1953), who saw it as a necessary complement to natural selection occurring later.

(pardon the pun) throughout the ecosystem (as in the dung beetles) or lineage (as in the dogs and cnidarians) in which the organism develops. Evo-devo (and eco-devo) demand contexts, either environmental or phylogenetical. A single organism is not a system. By constituting model systems as the genotype plus the spatial and temporal contexts of the organism we might be able to rehabilitate the notion of model system and use it constructively and creatively for evolutionary developmental biology.

I've had a longtime interest with the extrapolations between the body and the body politic (e.g. Gilbert 1979, 1997). It seems to me that there is a close parallelism between the model systems for the developing corporal body and models of the developing political body. There are three major models of the body politic: realist, liberal, and constructivist (see Walt 1998; Copeland 2000). The genetic model systems in developmental biology approximate the "realistic" view of the body politic. Central authority is the major determinant of the body's characteristics, and each individual, be it body or body politic, can be regarded as an entity with its own internal logic for survival and maintenance. The genes are the central authorities running the show, and it's a zero-sum game. The embryological model systems of developmental biology are like the "liberal" view of the body politic. Here, politics is not a zero-sum game, but a situation where win-win strategies are possible. There are several centers of authority, and interactions between these centers make possible unique and emergent institutions. Last, the evo-devo/eco-devo model systems of developmental biology model the "constructivist" model of the body politic. Here, the body is generated not only by internal factors but by the interactions of that body with its biotic, cultural, historical, and environmental environment. That is to say, the body's historical and situational contexts are critical here, and globalization is the norm.

Rowland Davis (2004) has suggested that the importance of model systems is waning, and that systems biology and comparative genomics have brought back diversity to biology and have downplayed the importance of individual model systems. His critique parallels the critique that social philosophers (see Martin 1992; Gilbert 1997) have made concerning the "death of the body." Emily Martin (1992) claimed, "We are not seeing the end of the body, but rather the end of one kind of body and the beginning of another kind of body." She contended that the very boundaries of the body and the social and biological agents that constructed it had changed from being an internal dialogue to one between internal and environmental components. These arguments are now being made in developmental biology, as well (McFall-Ngai 2002; Gilbert 2003c). What we are seeing is not the "end of model bodies" but the beginning of a new type of model body. We are seeing the replacement of "realist" and "liberal" models of body development by "constructivist" models. The new model systems not only reflect a new approach to developmental biology; they also reflect a new model of reality.

Acknowledgments This work was made possible through NSF grant RUI 0748508. The author wishes to thank Leah S. Gilbert for her discussions on political models.

Chapter 6
Niche Construction in Evolution, Ecosystems and Developmental Biology

John Odling-Smee

6.1 Introduction

In a volume devoted to re-examining our capacity to explain recent developments in the life sciences, the concept of niche construction should be able to make a useful contribution. If it cannot, there is likely to be something wrong with the concept; but I think it can.

Explaining anything in biology ultimately invokes evolutionary theory. The 20th century saw major advances in evolutionary theory, leading to more powerful explanations in many areas in biology. In spite of these successes it is still surprisingly difficult to connect some areas in biology with evolution. One stubborn problem is how to link ecosystem ecology to evolution (O'Neill et al., 1986; Jones and Lawton, 1995; Holt, 2005). Another is how to integrate developmental biology to evolutionary theory (Schlichting and Pigliucci, 1998; West-Eberhard, 2003). A third is how to relate human cultural processes to human genetic processes correctly, without introducing distortions that are potentially politically dangerous (Feldman and Laland, 1996; Odling-Smee et al., 2003; Richerson and Boyd, 2005). Why has evolutionary theory been so successful in supporting progress in some areas in biology, yet far less successful in others?

The paradox is probably no accident. Due to the way evolutionary theory was originally set up, the standard theory of evolution neglects a major causal process in evolution, now called niche construction (Lewontin, 1983a; Odling-Smee et al., 2003). The omission of niche construction meant that the standard theory of evolution was incomplete, and its incompleteness has subsequently disconnected evolutionary theory from other important areas in biology. I shall consider two of them here, ecosystem ecology, and developmental biology.

J. Odling-Smee (✉)
Institute of Biological Anthropology, University of Oxford, Oxford, UK;
Director of Studies & Lecturer in Human Sciences – Mansfield College, Oxford, UK
e-mail: john.odling-smee@anthro.ox.ac.uk

A. Barberousse et al. (eds.), *Mapping the Future of Biology*, Boston Studies in the Philosophy of Science 266, DOI 10.1007/978-1-4020-9636-5_6,

6.2 Niche Construction

Figure 6.1a summarises how evolution is currently thought to work. Autonomous natural selection pressures in environments (E) act on populations of diverse organisms (or phenotypes) to influence which individuals survive and reproduce, and pass on their genes to the next generation via genetic inheritance. Consequently, the adaptations of organisms are assumed to be the selected products of natural selection moulding organisms to fit pre-established environmental templates. The templates are dynamic because processes that are independent of organisms frequently change the worlds to which organisms have to adapt. However, the changes that organisms bring about in their own worlds are seldom thought to have evolutionary significance, and they are certainly not regarded as an evolutionary process in their own right. Williams (1992b, p. 484) was succinct:

> Adaptation is always asymmetrical; organisms adapt to their environment, never vice versa.

The problem with this view is that it discourages consideration of the feedback in evolution caused by the modification of environmental natural selection pressures by the actions of organisms. Organisms, through their metabolisms, behaviours and choices, partly define, create and destroy their own selective environments (Lewontin, 1983a; Jones et al., 1994, 1997; Wright and Jones, 2006. Boogert et al., 2006). In doing so they transform some of the natural selection pressures that act on themselves, and on each other. Subsequently, these transformed selection pressures feed back to select for different genotypes in populations from those that would have been selected in the absence of this feedback (Laland et al., 1996, 1999). Hence the adaptations of organisms cannot be exclusively consequences of organisms responding to autonomous natural selection pressures in their environments. Sometimes they must involve organisms responding to natural selection pressures that have previously been transformed either by their own actions, or by the actions of their ancestors, or by the actions of other organisms, in shared ecosystems.

Lewontin (1983a) illustrated this point by using two pairs of deceptively simple differential equations. Standard evolutionary theory, he said, can be summarised by Equations 6.1 and 6.2.

$$dO/dt = f(O, E), \tag{6.1}$$

$$dE/dt = g(E). \tag{6.2}$$

In Equation 6.1 evolutionary change in organisms during time, dO/dt, depends on both organisms states, O, and environment states, E. In Equation 6.2, however, environmental change during time, dE/dt, depends only on environmental variables (E). It does not depend on the actions of organisms, O. This is because adapted organisms are not thought to cause any of the environmental changes that subsequently select for adapted organisms.

But, said Lewontin, that's not how evolution works. How he thinks it does work is summarised by Equations 6.3 and 6.4.

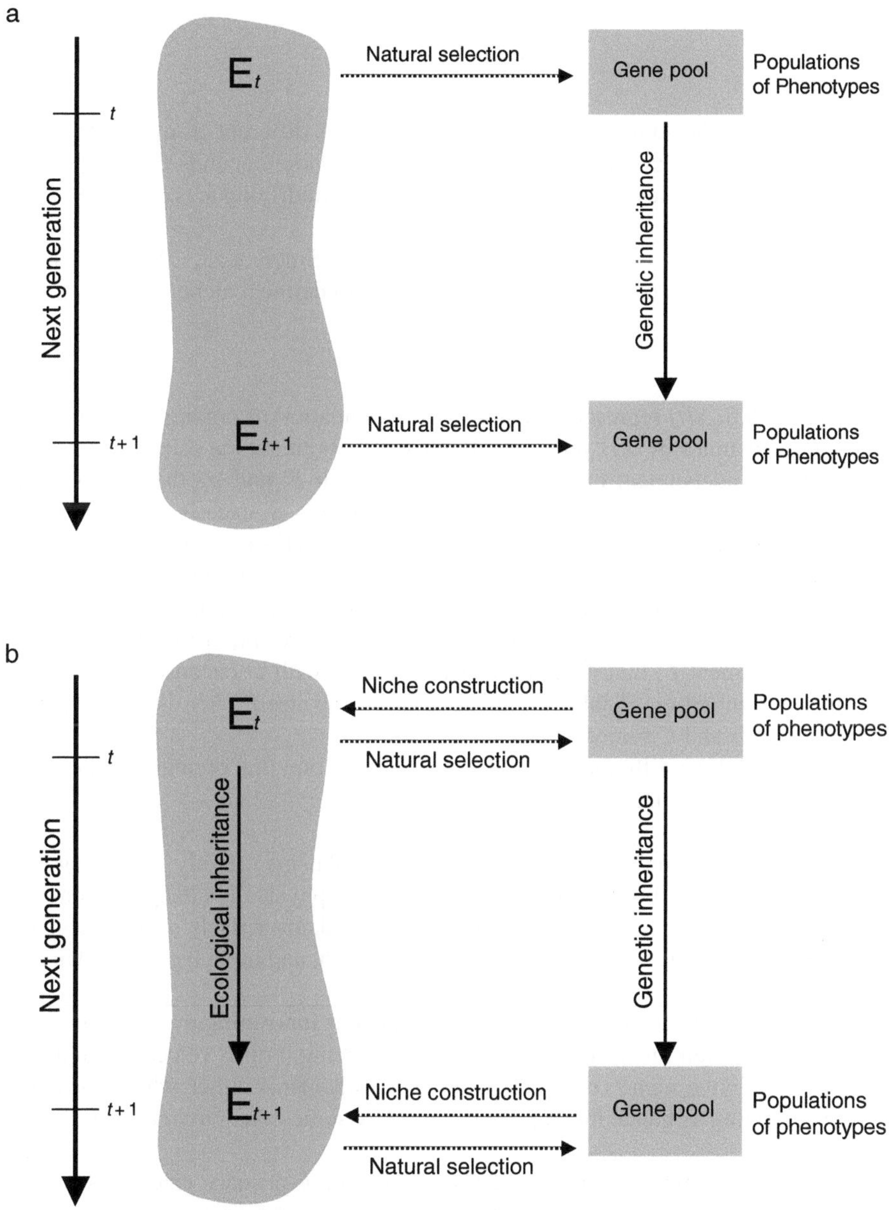

Fig. 6.1 (a) Standard evolutionary theory (b) Extended evolutionary theory

$$dO/dt = f(O, E), \tag{6.3}$$

$$dE/dt = g(O, E). \tag{6.4}$$

Equation 6.3 is the same as 6.1, but Equation 6.4 is different from 6.2. In Equation 6.4, environmental change during time, dE/dt, now depends on independent environmental variables, E, and on the environment modifying acts of active organisms, O.

Lewontin's position is very close to ours (Odling-Smee et al., 2003), so I'll use it to introduce niche construction. The first step is to define a niche.

$$N(t) = h(O, E) \tag{6.5}$$

In Equation 6.5, *N(t)* represents the niche of a population of organisms O at time t, where the dynamics of *N(t)*, are explicitly driven by both natural selection pressures arising from independent environmental variables in *E*, and by the environment-modifying "niche-constructing" activities of organisms *O*, exactly as described by Lewontin's Equations 6.3 and 6.4. In our definition, *N(t)* is an evolutionary, as well as an ecological niche. Everything in Equation 6.5 is evolving. The population, *O* is evolving, as usual. *O's* selective environment, *E*, is in part coevolving as a consequence of *O's* genetically "informed", and possibly "brain-informed" "work" on its environment *E*, that is to say it is coevolving with *O* because of *O's* niche-constructing activities. Finally the niche relationship itself, *N(t)*, is evolving as a function of *O's* and *E's* interactions.

We classified the different kinds of niche construction that organisms can potentially express. The main kinds are these:

(i) Perturbational niche construction involves organisms actively modifying their environments at particular times and places, by physically changing them.
(ii) Relocational niche construction[1] involves organisms actively moving in space, and choosing the direction or distance they travel, and often the time when they travel as well.
(iii) Inceptive niche construction involves organisms innovating by initiating a new change in their environments, either by perturbing it or by relocating in it.
(iv) Counteractive niche construction involves organisms either wholly or partly neutralising a change that has been caused by some other environmental agent.

We also superimpose a further distinction on these primary categories by distinguishing between niche construction that only affects an organism's "private world", or its focal environment, as opposed to niche construction that affects other organisms too, in shared ecosystems. Finally, we distinguish between positive and

[1] Relocational niche construction need not change any component in O's environment directly. In contrast to perturbation, it does not promote "environmental construction". What it changes is the relationship between O and E, and thus O's niche. It is in this sense that relocation is another kind of niche construction.

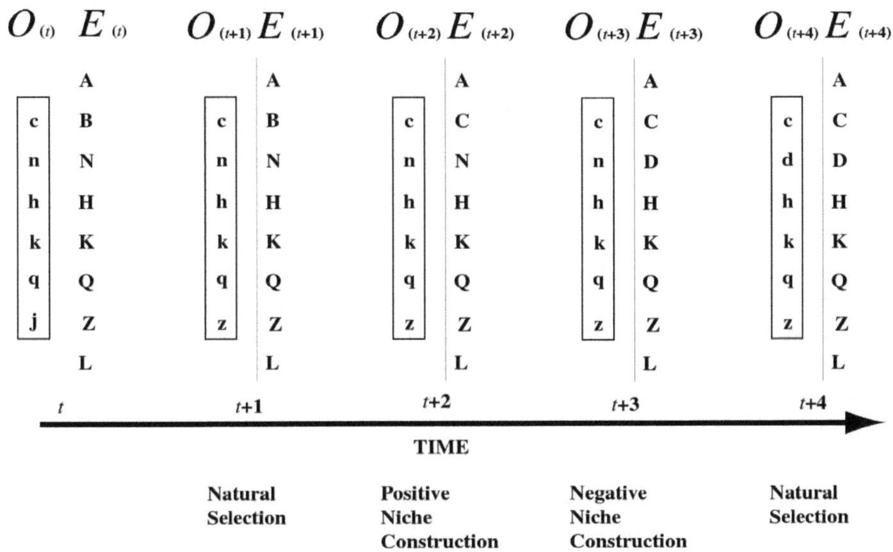

Fig. 6.2 A short sequence in the evolution of a niche

negative niche construction. Positive niche construction refers to acts that increase the fitness of the niche-constructing organisms. Negative niche construction refers to acts that decrease their fitness. Most niche-construction is positive in the short term, but in the longer term it may become negative, through the over-exploitation of a resource, for example, or possibly, through pollution.

Figure 6.2, based on Bock (1980) and Odling-Smee et al. (2003), illustrates a short sequence of possible "niche evolution". Here, traits of organisms are classified as sets of features, symbolised by the lower case letters in *O*, and natural selection pressures in their environments are classified as sets of factors, symbolized by the upper case letters in *E*. The [*OE*] niche relationship between *O* and *E* is assumed to be adaptive whenever there is a "match" between the lower and upper case letters, but it is not adaptive whenever there is a "mis-match".

At time t, at the start of this sequence, *O* is already tolerably well adapted to its environment, *E*, because most of the lower and upper case letters match.

But there are two mismatches, between b and C, and between j and Z. At time t + 1, one of these mismatches disappears. Natural selection pressures, due to factor Z in the environment, have now selected for feature z in *O* instead of feature j. So *O* is now better adapted to its environment than it was before, thanks to natural selection, as described by standard evolutionary theory.

The next step, however, is not standard. At time t + 2, *O* has improved its own adaptation still further by positive niche construction. *O* has changed factor B in its environment to factor C without changing itself. Feature c in *O* is still c. So this time *O* has achieved a new feature-factor adaptive match by causing a change in its environment to suit itself, for example by digging a burrow, rather than by responding to a natural selecting pressure in its environment with a change to itself.

At time t + 3, *O* damages its environment by negative niche construction, possibly by polluting it. *O* thereby causes factor N in its environment to change to D. That then generates a new mismatch, between n and D, that is subsequently corrected at time t + 4, by natural selection again, as per standard evolutionary theory.

6.3 Earthworms

We need some examples. Elsewhere, we have given numerous examples of niche construction from all the kingdoms of life (Odling-Smee et al., 2003). Space is limited, so I will describe only one example, earthworms, in some detail, and then a few others more briefly.

Earthworms have the attraction of being studied by Darwin. In a book he wrote towards the end of his life, Darwin (1881) pointed out that earthworms are terrific niche-constructors, although he didn't use that term. By burrowing, dragging organic material into the soil and mixing it up with inorganic material, and by their casting, earthworms dramatically change both the structure and chemistry of soils. Soils that contain earthworms typically demonstrate enhanced plant yield, less surface litter, more topsoil, more organic carbon, nitrogen and polysaccharides, as well as better porosity, aeration and drainage (Satchell, 1983; Lee, 1985). So earthworms must live in partly self-constructed worlds. Consequently, at least some of the natural selection pressures that act on earthworms must have been transformed by the prior activities of many generations of earthworms.

One evolutionary consequence of these soil-changing activities of earthworms was studied by Turner (2000). Turner compared the physiological characteristics of earthworms, in particular the characteristics of their kidneys, with those of other animals, and he made a surprising discovery. Earthworms appear to be equipped with the "wrong" kidneys. Different kinds of kidneys are typically found in different animals, depending on where they live. Turner compared freshwater, marine and terrestrial kidneys and he pointed out that each kind of kidney does a different adaptive job. Animals living in freshwater are in danger of being flooded by excess water, so freshwater kidneys have to excrete surplus water. Animals living in the sea are in danger of being killed by excess salts, so marine kidneys have to get rid of salts. Animals living on land are in danger of drying up, so terrestrial kidneys must prevent desiccation by retaining as much water as possible. The crucial point is that earthworms, which are obviously terrestrial animals, are equipped with very nearly typical freshwater kidneys. What seems to have happened is that earthworms have retained the freshwater physiology of their earlier aquatic ancestors, even though they now live on land. These originally aquatic creatures appear to have solved their water- and salt-balance problems through niche construction, for instance, by tunnelling, exuding mucus, and eliminating calcite. The well-aggregated soils they produce weaken matric potentials, and make it easier for them to draw water into their bodies (Turner, 2000). All of this earthworm activity highlights a problem with the concept of adaptation. In this case it is the soil that does the changing, rather than the worm, to meet the demands of the worm's freshwater physiology.

What earthworms seem to be telling us is that adaptation is a two-way process. Adaptation is not just a matter of natural selection selecting for adaptive traits in responsive organisms, as proposed by Williams (1992b). It is also a matter of organisms transforming at least some of the natural selection pressures they encounter in their local environments to suit themselves, and/or succeeding generations, by niche construction, as proposed by Lewontin (1983a).

6.4 Other Examples

I'll give other examples of niche construction more rapidly. They are mostly taken from recent empirical papers. One concerns forest fires. Flammability in forests can be a "niche construction trait" (Schwilk, 2003). Several species of plants, particularly pines, and chapparal species, increase the frequency of forest fires by allowing their oils or litter to accumulate to the point where a fire becomes inevitable. One consequence of this kind of niche construction is that such species are forced by their own environment-modifying acts to evolve fire resistance, as well as efficient resprouting following fires. They gain by out-competing less fire resistant competitors. For instance, in some species of pines, the pines require a fire to suppress competing plants before their seeds will germinate.

A second plant example comes from *Arabidopsis thaliana*, a plant that botanists use as a model organism. Donohue (2005) describes how niche construction in plants frequently occurs through plasticity in developmental phenologies, and she showed how the environment created by seed dispersal in *Arabidopsis thaliana* can cause novel, plasticity-induced genetic constraints on the subsequent evolution of seed dispersal. As seed dispersal evolves, the environment experienced by seeds changes in ways that reduce the expression of genetic variation, and impede further evolutionary responses to selection. Donohue also showed how interactions between two other "niche-constructing traits" in *Arabidopsis thaliana*, flowering time and germination time, led to a novel bivoltine life history in which generation time was halved.

An intriguing animal example involves both plants and ants. Frederickson et al. (2005) became interested in so-called "devils" gardens' in the Amazonian rainforest. These "gardens" consist almost entirely of a single species of tree *Duroia hirsute*, even though they are surrounded by multiple competing plant species. According to local legend the gardens are cultivated by evil forest spirits. They are actually cultivated by ants (*Myrmelachista Schumani*). The ants nest in the stems of *D. hirsuta*, and create "devil's gardens" by poisoning all the other competing plants in the vicinity, by injecting formic acid into their leaves. In this example the ants' niche-constructing behaviour sets up a mutualism between the ants and their host trees. The resulting "gardens" can be long lasting. The largest devil's garden in Frederickson et al. study area was estimated to be 807 years old, so it must have been "inherited" by multiple generations of ants.

A different kind of animal example focused on "social niche construction" in a primate species, pigtailed macaques (*Macaca nemestrina*). Flack et al. (2006)

demonstrated that stable social networks in pigtailed macaques depend on "policing" by a small subset of high-status individuals who prevent conflicts from escalating by actively intervening and terminating them, or sometimes, just by their physical presence. If this "policing" is artificially disabled by the "policers" being physically removed by an experimenter, then the "social niche" destabilizes and social disorder increases. Flack et al. reported that "policing" can significantly alter the construction of the social resource networks that make group living advantageous. For example, they noted that when "policing" was operational, group members built larger social networks that were characterized by greater partner diversity, and increased socially positive cooperation.

There are also examples of human "cultural niche construction", by which is meant niche-construction that is generated by human cultural activities. The best known is probably dairy farming. Human pastoralists typically go on drinking milk from their cattle as adults, and by doing so for many generations, they have apparently modified a natural selection pressure in their environments in favour of a gene that increases the ability of human adults to absorb lactose. This ability allows people to continue to drink milk long after their infancy is over. In contrast, human adults from non-pastoralist areas usually avoid milk because it can make them sick. Thus dairy farmers appear to have affected their own genetic evolution by "agricultural" niche construction (Cavalli-Sforza and Feldman, 1981; Durham, 1991; Enattah et al., 2002). Other examples of recently modified natural selection pressures, selecting for new alleles in human populations as a consequence of diverse kinds of cultural niche-construction, are emerging from new insights into the human genome (Voigt et al., 2006; Wang et al., 2006).

6.5 The Limitations of the Standard Theory

Many examples of niche construction are already well known to biologists, and the fact that organisms can and often do alter their environments is known by all ecologists (Odling-Smee et al., 2003). So how does standard evolutionary theory deal with these phenomena? Is it really necessary to go beyond standard evolutionary models to incorporate niche construction in evolution?

Many instances of niche construction are described by models broadly compatible with standard evolutionary theory. For example, niche-construction has been captured by models of frequency- and density-dependent selection (Futuyma, 1998); habitat selection (Hanski and Singer, 2001); maternal inheritance (Kirkpatrick and Lande, 1989); extended phenotypes (Dawkins, 1982); indirect genetic effects and epistasis (Wolf, 2000); and co-evolution (Thompson, 1994). In spite of these approaches the full significance of niche construction has been neglected. What has been missing is a body of theory that explicitly recognises niche construction as a co-directing cause in evolution working in conjunction with natural selection (Waddington, 1969; Lewontin, 1983a; Odling-Smee et al., 2003). Why has such a seemingly obvious process been marginalised for so long by standard evolutionary theory?

Probably the answer lies in a seldom re-considered foundation assumption of standard evolutionary theory concerning the role of environments in evolution. Godfrey-Smith (1996) drew attention to it by describing standard evolutionary theory as an "externalist" theory of evolution. By "externalist" Godfrey-Smith meant that standard evolutionary theory seeks to "explain" the internal properties of organisms in terms of the external properties of their environments. Consequently, the adaptations of organisms are explained exclusively in terms of natural selection pressures in their external environments (Equations 6.1 and 6.2).

There is a detectable Newtonian residue concealed in this assumption. The view of organisms as: ".... passive objects moulded by the external force of natural selection" (Lewontin, 1983, p. 275) recalls Newtonian physics. A "force", natural selection, acts on a "body", a population of passive organisms that "reacts" by evolving. Standard evolutionary theory may never have been fully weaned from Newton.

The principal point that the "externalism" of standard evolutionary theory obscures is that organisms are not "passive objects" (Waddington, 1969; Lewontin, 1983a). They are alive, and active, as well as reactive, and unlike non-living objects, they can "push back". To stay alive organisms have to gain resources from their external environments by non-random "work". Non-random work requires organisms to be "informed" by "meaningful information", minimally by naturally selected information carried in their genes. "Informed" organisms are bound to impose some non-random changes on their environments by their "informed" work (Odling-Smee et al., 2003). Hence, informed active organisms are compelled not only to respond to natural selection, but also to "choose" and "perturb" specific components of their environments, and therefore to change some of the selection pressures that select them.

Let us reconsider two putative "causal arrows" of evolution in the light of Godfrey-Smith's (1996) critique. One "arrow", natural selection, points from environments to organisms. The second "arrow" niche construction, points in the opposite direction, from organisms to environments. Given the direction in which it is pointing, the natural selection "arrow" is clearly compatible with the "externalist" assumption of standard evolutionary theory. It is conceptually straightforward to describe how the external properties of environments, in the guise of natural selection, can "cause" the internal properties of organisms by selecting for adaptations. In contrast the niche construction "arrow" is, equally clearly, not compatible with the "externalist" assumption of the standard theory because it is pointing in the "wrong" direction. That makes it difficult or impossible to describe any changes in natural selection that are caused by niche-construction as evolutionarily "causal". Instead, the standard theory is repeatedly forced by its own assumption to "write off" observed instances of niche construction as nothing more than phenotypic, or possibly extended phenotypic (Dawkins, 1982) consequences of prior natural selection, and to model them accordingly, as consequences, but never causes of evolution.

In spite of this biasing assumption, many biologists do now accept that niche construction is a causal process in evolution, but some are still trying to reconcile niche construction with standard evolutionary theory by claiming it is only locally causal. This amounts to accepting two unequal causal arrows in evolution. One "cause", natural selection is "universal" and potent. The other "cause", niche construction, is local and weak.

Odling-Smee et al. (2003) anticipated this criticism by describing the "universal" properties of niche construction (Table 4.1, p. 176), partly on the basis of the probabilistic physics derived from the Maxwell's Demon concept (Bennet, 1987; Jaynes, 1996). We claimed that natural selection and niche construction are approximately equally potent causes in evolution. Life would be impossible without either process, as "origin-of-life" researchers know (Fry, 2000). We also claimed that niche construction is a developmental process that is capable of co-directing the evolution of populations via a Darwinian, as opposed to a Lamarckian, mechanism. Therefore niche construction should provide a powerful "evo-devo" link. I'll rehearse the arguments.

To stay alive and reproduce without violating the second law of thermodynamics, all organisms have to "construct" coupled [OE] systems, or "niches", that constantly enable them to exchange energy and matter with their local environments. Organisms must live at the "expense" of their environments, by consuming energy-rich sources of free energy, and by exporting detritus. The construction of these [OE] niches has two consequences for evolution.

First, by taking energy and matter resources from their environments to survive and reproduce, organisms supply the "raw material" for natural selection to sort, in the form of successive generations of diverse organisms in populations. So niche construction has the effect of "fuelling" evolution by constantly gaining energy and material resources from environments, and by turning these resources into offspring (Odling-Smee et al., 2003). This consequence is standard. So far it has failed to integrate developmental and evolutionary biology sufficiently to explain a great deal of now rapidly accumulating data, particularly molecular data (Gilbert, 2003c). Instead, it continues to permit developmental biology, inclusive of niche construction, to be disconnected from evolutionary biology through the sequestering of the germ line (West-Eberhard, 2003; Amundson, 2005; Jablonka and Lamb, 2005),

The second consequence of niche construction, however, cannot be disconnected from evolution because it depends on interactions between both of evolution's "causal arrows", the "fuelling" process of niche construction, and the "informing" process of natural selection. These two processes interact because obligate exchanges of energy and matter between organisms and environments, in coupled [OE] systems, necessarily affect environments as well as organisms. As Lewontin (1983a) pointed out, organisms do not evolve relative to independent environments (Equations 6.1 and 6.2). In part, they co-evolve with their environments (Equations 6.3 and 6.4). In an organism-environment co-evolutionary scenario the niche-constructing actions of organisms inevitably modify some sources of natural selection in their own, and each other's environments. That leads to the selection of different genes, in diverse organisms, in diverse populations, and turns niche construction into a second "causal arrow" in evolution.

This niche constructing "arrow" is not merely locally "causal". It is potentially as general a cause in evolution as is natural selection itself. The environmental consequences of niche construction often accumulate, and when they do they can influence complex ecological inheritance systems for multiple populations in diverse ecosystems (Odling-Smee et al., 2003). The scale of these accumulations varies enormously, from extremely local, to very general, and it can become global. A well-known global example is the replacement of the earth's early anaerobic atmosphere by its present aerobic atmosphere, probably as the cumulative consequence of the niche-constructing activities of billions of photosynthesising organisms, over millions of years (Odum, 1989; Holland, 1995). Subsequently, the earth's "niche constructed" aerobic atmosphere probably affected the evolution of countless species, and hence the historical course of evolution itself.

The global consequences of niche construction were implicitly recognised by Dietrich and Taylor Perron (2006). These authors used a number of biotic effects on planet earth as their base line for a search for topographic signatures of life on other planets. They have not yet found any convincing signatures on any other planet, but their work is an intriguing application of niche construction. Again, it emphasises that the consequences of niche construction can be far from local. They can occur on a planetary scale.

6.6 Extended Evolutionary Theory

We may now draw two conclusions. First, there are two "causal arrows" in evolution, natural selection and niche construction. Second, it will not be possible to include both these "causal arrows" in a comprehensive theory of evolution without dropping the externalist assumption of standard evolutionary theory.

What is required is an "interactionist" theory of evolution that is not biased *a priori* by any foundation assumption, either in favour of natural selection and against niche construction, or in favour of niche construction and against natural selection. It must recognise that both the internal properties of organisms, and at least some properties of their external environments, are caused by the mutual interactions of niche-constructing organisms and their naturally selecting environments. For this reason, when we initially proposed extending the theory of evolution to incorporate niche construction, we had to start by re-orienting the theory around the "interactionist" concept of a "niche", instead of basing it on the "externalist" assumption of an autonomous external environment (Odling-Smee et al., 2003). The niche concept, N(t), works in this role because it is theoretically "neutral". It is bias-free, and it permits evolution to be understood in terms of the interactions of both niche-constructing organisms and their naturally selecting environments (Odling-Smee, 1988). Niches are "constructed" by organism -environment interactions, as in Equations 6.3, 6.4, and 6.5.

The resulting "extended theory of evolution" is summarized in Fig. 6.1b. This figure illustrates a scheme in which the evolution of organisms now depends on both natural selection and niche construction. Genes are transmitted by ancestral organisms to their descendents as directed by the outcomes of natural selection, exactly as in Fig. 6.1a. However, selected habitats, modified habitats, and modified sources of natural selection in those habitats are also passed on by these same organisms to their descendents, via an "ecological inheritance", under the direction of niche construction. Hence, the selective environments encountered by organisms are partly determined by independent sources of natural selection, exactly as described by the standard theory, for instance, by climate, weather, of physical and chemical events. But they are also partly determined by what organisms do, or previously did to their own and each others' environments, by niche construction.

Ecological inheritance (Fig. 6.1b) is a second putative new process in evolution. It refers to the inheritance via an external environment of one or more natural selection pressures that have previously been modified by niche-construction, and it works differently from genetic inheritance (Odling-Smee, 1988, 2003). For instance, unlike genetic inheritance, ecological inheritance does not depend on the transmission of discrete "replicators" (Dawkins, 1976). It only depends on the persistence between generations of whatever physical changes are caused by the niche-constructing activities of ancestral organisms in the local selective environments of their descendents. A second difference is that when organisms inherit naturally selected genes, they inherit genetically encoded semantic information (Odling-Smee et al., 2003). But when organisms inherit modified natural selection pressures via ecological inheritance they do not inherit any information directly. Instead, they inherit transformed natural selection pressures in their environments. Subsequently, these transformed selection pressures select for different genes, and by doing so, determine the "content" or "meaning", or "fitness", of whatever semantic information is encoded in these different genes. A third difference is that in sexually reproducing organisms genetic inheritance is transmitted by only two parents, and from the point of view of an offspring organism, on a single occasion only. In contrast, ecological inheritance is transmitted by multiple organisms throughout their lives.

6.7 Modelling Niche Construction

We modelled niche construction by using two-locus population genetic models (Laland et al., 1996, 1999, 2001). I will suppress the maths, but will briefly describe the logic that underpins all our models. We focussed on a single population, and on two genetic loci only, which we labelled **E** and **A**. We then assumed that: (i) The population's capacity for niche construction is influenced by the frequency of its alleles at the first, or **E** locus. (ii) The amount of some resource, **R**, in the population's environment, depends either wholly, or in part, on the niche-constructing activities of past and present generations of organisms. (iii) The amount of this resource, **R**, subsequently influences the pattern and strength of selection acting on alleles at the second, or **A** locus.

These assumptions need some qualifying. First, the assumption that niche construction is influenced by the **E** locus does not mean that niche construction is always the expression of a single gene, but only that, to keep things simple, we initially focused on only one of the many genes that might influence niche construction. Second, the resource **R** could be any environmental resource that is modified by niche construction. It could be biotic, or abiotic. For instance, it might be a food item, a parasite, a predator, or a water resource, detritus, a chemical element in the soil, or even an inherited artefact. It only matters that **R** is both a source of natural selection on the **A** locus, and that it is modified by the niche constructing activities of a population's phenotypes under the influence of **E** locus genotypes. It is also possible for changes in **R** to depend only partly on the focal niche constructing population. They might also depend, in part, on other independent agents in the environment, including abiotic agents (Laland et al., 1999). Third, this basic model can be generalised because the **A** locus need not always be in the same population as the **E** locus. The two loci could be in different populations as occurs, for instance, in co-evolving populations.

The results we obtained demonstrated that adding niche construction to evolutionary models can make a lot of difference. For example, we found that modified natural selection pressures due to niche construction can override independent sources of natural selection, and drive populations along evolutionary trajectories that they would not have taken if they had not expressed the niche construction. It is also possible for niche construction to initiate new evolutionary episodes by changing selective environments that would otherwise have remained unchanged. Niche construction can also influence the amount of genetic variation in a population by affecting the stability of polymorphic equilibria, and by generating new equilibria, and it can generate unusual evolutionary dynamics, for instance, by causing timelags, and momentum and inertia effects. Finally, niche construction can reverse an evolutionary response in a population. Our general conclusion is that adding niche construction and genetic inheritance to evolutionary models enhances our understanding of how evolution works.

6.8 The Implications of Niche Construction for Ecology

We can now return to those major areas in biology that have previously been difficult to integrate with standard evolutionary theory. Does substituting extended evolutionary theory (Fig. 6.1b) for standard evolutionary theory (Fig. 6.1a) ease any these difficulties? Let us consider ecosystem ecology first.

One point of contact between niche construction and ecosystem ecology can be made immediately. Jones et al. (1994, 1997); Jones and Lawton (1995); Gurney and Lawton (1996), and Wright and Jones (2006) call organisms that choose and perturb components of their environments "ecosystem engineers", and "ecosystem engineering" is essentially the same thing as niche-construction. The only major difference is that we are primarily interested in the evolutionary consequences of niche construction, whereas this group of ecologists is primarily interested in the ecological consequences of ecosystem engineering. So how does the explicit inclusion of niche construction and ecological inheritance in evolution affect the relationship between ecosystem ecology and evolutionary theory?

6.9 A Simple Model of an Ecosystem

We need a simple model of an ecosystem. Ecosystems typically include diverse species of organisms, and diverse abiota. For our purpose, however, it is sufficient to reduce all of them to just two "elementary kinds" of ecosystem components, living organisms, or biota, and non-living abiota. We also need a set of "elementary links" to connect these elementary ecosystem components to each other. We assume that a "link" occurs whenever a "source" ecosystem component, which could be either biotic or abiotic, emits any kind of output, that subsequently becomes an input for any kind of "sink" component in an ecosystem. We also assume that every ecosystem component emits at least one output(o/p), and receives at least one input(i/p), from some other ecosystem component (Fig. 6.3).

To make these concepts concrete it may help to give a familiar example, Suppose a "source" ecosystem component is a population of cows, and that the "outputs" from these cows are cowpats. Then the cowpats, including all the chemicals they contain, will also be "inputs" for a "sink" component, the soil on which they fall, and possibly for other "sink" components too, for example, for a population of dung-beetles for which the cowpats are a valuable resource. So there is a "link" between a biotic "source", the cows and an abiotic "sink" the soil, and possibly a second "link" too between the same biotic "source", the cows again, and a biotic "sink", the dung-beetles.

When these two elementary kinds of ecosystem components, biota and abiota, interact with each other, they yield a comprehensive set of four possible kinds of elementary links between them. These are: (i) biotic → biotic links (for example, predator-prey or host-parasite interactions); (ii) abiotic → biotic links (for example, the uptake of water by a population of plants); (iii) biotic → abiotic links (for example, the production of oxygen by photosynthesising plants), and (iv) abiotic → abiotic links (for example, chemical reactions). We assume that all webs and chains in all ecosystems are made up of assemblies of these four elementary kinds of interactive links, and that the resulting webs and chains of connection among both biotic and abiotic ecosystem components are largely responsible for the structures and dynamics of ecosystems (Odling-Smee et al., 2003). Each of these elementary links is illustrated in Fig. 6.3a. In this figure living organisms, or biota, are symbolised by shaded circles, non-living abiota are symbolised by open squares, and the links between them are symbolised by connecting lines.

The immediate consequence of any interaction between any source and sink is a change of state in both the sources and sinks. For the moment we will disregard the changes of state that occur in source ecosystem components, and consider only those that occur in sinks. Following, any interaction, either an ecologically

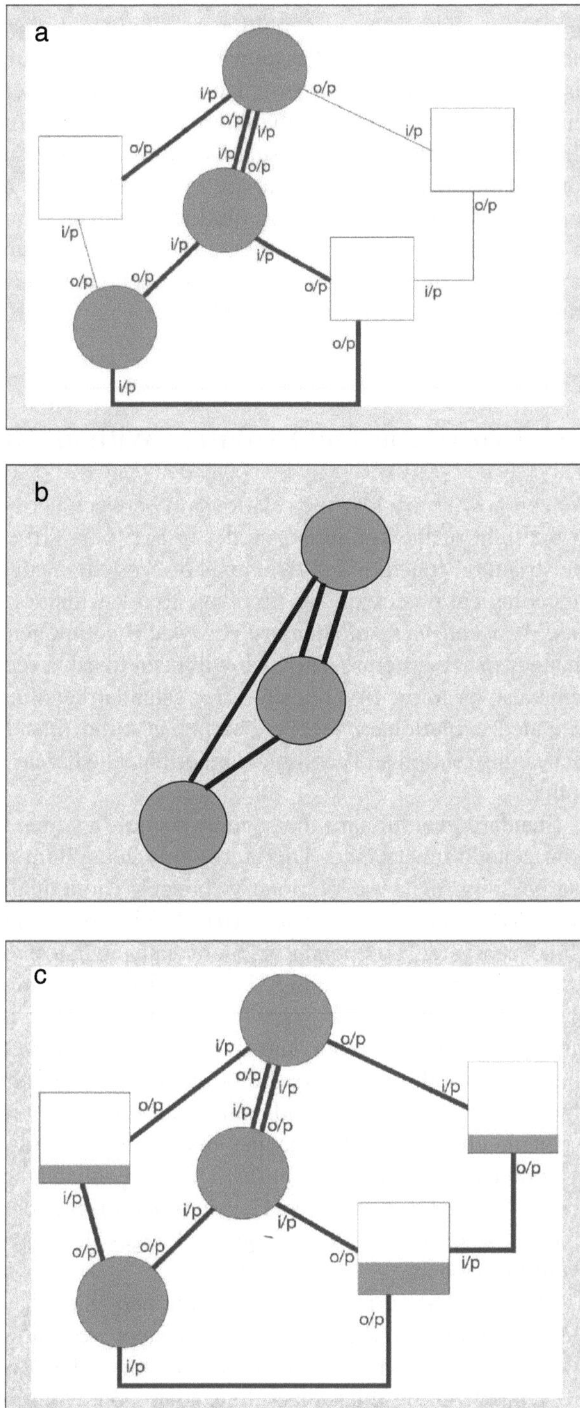

Fig. 6.3 (a) Webs of connectance in ecosystems, composed of a mix of evolutionary and non-evolutionary links (b) Ecosystems are reduced to (e.g.) food webs by the population-community approach (c) Webs of connectance in ecosystems after adding niche construction: All links are now evolutionary

significant change, or an evolutionary significant change, or both, could occur in the sink component. I will assume that ecological changes primarily involve energy and matter exchanges, for instance a flow of a nutrient between different ecosystem components. However, evolutionary changes primarily refer to changes in the relationship between natural selection pressures stemming from source ecosystem components, and the genes that are selected by those natural selection pressures in sink ecosystem components. Evolutionary changes are therefore "fitness" changes in ecosystem components. They are ultimately measured by a different "currency" from ecological changes, a fitness or a genetic currency, instead of energy or matter currencies (Odling-Smee et al., 2003).

6.10 Ecosystems and Evolution Without Niche Construction

We can now reconsider the relationship between ecosystem ecology and evolution in the light of the simplified model in Fig. 6.3a. To establish the extent to which the structure, function and dynamics of ecosystems depend on evolution as well as on ecological processes, we first have to ask which of these elementary ecosystem links between these elementary ecosystem components can mediate evolutionary changes in ecosystems? Then we will need to ask a second question. Do we get the same answer to the first question from standard evolutionary theory as we do from extended evolutionary theory, bearing in mind that the standard theory excludes niche construction and ecological inheritance, whereas the extended theory includes both?

Standard evolutionary theory only models a single kind of inheritance in evolution, genetic inheritance. Therefore, only ecosystem components that carry genes can possibly register evolutionary changes. Biota qualify, but abiota do not. Hence links (i), biotic → biotic and (ii) abiotic → biotic where, in both cases, the "sink" ecosystem components is biotic, can mediate evolutionary changes. Biota can respond to naturally selecting "inputs", from either biotic or abiotic "source" components with genetic changes, which can be described in "fitness" terms. In contrast, links (iii) biotic → abiotic and (iv) abiotic → abiotic cannot mediate evolutionary changes. In both these links the "sink" ecosystem component is abiotic. However, abiota do not carry genes and cannot respond to any input from any "source" with a genetic change. So the changes that occur in abiota cannot be measured by the evolutionary currency of "fitness". They can only be measured in terms of ecology's energy and matter currencies. Hence, links (iii) biotic → abiotic and (iv) abiotic → abiotic can only mediate ecological changes in ecosystems. They cannot mediate evolutionary ones.

Thus, according to standard evolutionary theory, the mix of all four kinds of elementary links in the miniature ecosystem web shown in Fig. 6.3a includes both evolutionary links and non-evolutionary links. The evolutionary links, (i) and (ii), are symbolized by thick black lines. The non-evolutionary links, (iii) and (iv) are symbolized by the thin black lines.

It is this mix of evolutionary and non-evolutionary links in ecosystems that has hitherto made it so difficult to fully integrate ecosystem ecology with standard evolutionary theory (Jones and Lawton, 1995; Odling-Smee et al., 2003). In practice it has typically forced ecologists to choose between two alternative approaches to ecosystems, the process-functional approach, versus the population-community approach, depending on what questions they are asking (O'Neill et al., 1986; DeAngelis, 1992; Holt, 2005). The process-functional approach is more general and can capture everything except evolution. For example, it can potentially model and study complete biogeochemical cycles in terms of energy and matter flows through both biota and abiota. However, it cannot capture the extent to which those energy and matter flows, may themselves be partly governed or influenced by the evolved, and possibly still evolving, adaptations and activities of organisms. Conversely, the population-community approach is able to describe the evolution of biota in ecosystems, but only at the price of largely ignoring non-evolving abiota. For instance, a population-community approach to the miniature ecosystem in Fig. 6.3a, would probably reduce it to the food web illustrated in Fig. 6.3b before studying the evolution or co-evolution of its constituent populations.

6.11 Ecosystems and Evolution with Niche Construction

What happens when we add niche construction to evolution? Because niche construction can generate a second inheritance system in evolution, ecological inheritance, in addition to genetic inheritance, it immediately provides ecologists with an extra inheritance system to work with. This step is significant. Although the abiotic components of ecosystems do not carry genes, and cannot transmit a genetic inheritance between successive generations of organisms, what they can do is respond to niche-constructing organisms, and transmit modified natural selection pressures that are caused by ancestral niche-constructing organisms in abiota, to descendant organisms, in the form of an ecological inheritance. So, even though abiotic ecosystem components cannot participate in evolution via genetic inheritance, they can participate in evolution via ecological inheritance

I'll illustrate that by using a specific example: the earthworms again, and I'll initially concentrate on link (iii), that is on the biotic → abiotic links illustrated in Fig. 6.3a. According to standard evolutionary theory, link (iii) is a non-evolutionary link because the "sink" ecosystem component is abiotic, and it cannot respond to any input from a biotic "source" with a genetic change. It can only respond with an ecological change. However, according to extended evolutionary theory, any ecological change could potentially contribute to an ecological inheritance for one or more evolving populations. For example, suppose in Fig. 6.3a, the biotic source in a link (iii) interaction is a population of niche-constructing earthworms, and that the worms change the physical state of the soil in which they live by their casting, burrowing, dragging leaves into the soil, and so on, then the effect on the soil should be a change in its physical state that is directly caused by the adaptations and niche-constructing activities of the worms.

I've symbolised these kinds of organism-imposed changes on abiota in Fig. 6.3c, by dividing the abiotic "sinks" in link (iii) interactions into two parts. In the example interaction we are considering, between the earthworms and the soil, the white part represents the physical state of the soil in the absence of niche construction, or the "null state" of the soil. The shaded part represents the extent to which the soil has been driven out of its "null state" by the genetically informed "work" of the niche-constructing earthworms. So the shaded part represents what could subsequently become an "ecological inheritance", in the form of one or more biotically modified natural selection pressures, for any population that subsequently used the soil. In this way, any abiotic sink, in any link (iii) interaction, potentially becomes an evolutionary, as opposed to a non-evolutionary link, in any ecosystem (compare Figs. 6.3a and 6.3c).

Whether a potential "ecological inheritance" actually becomes evolutionarily significant depends on what happens next. In the earthworm→ soil example, one possibility is that the change of state in the soil becomes evolutionarily relevant to later generations of the earthworms themselves, leading to the kind of consequences discussed by Turner (2000). Another possibility, and the more relevant one here, is that the changes in the soil caused by the niche-constructing earthworms could subsequently influence the evolution of a second population, for example, a population of plants that is growing in the soil, by serving as the source of a modified natural selection pressure in the soil, for the plants, this time via a subsequent link (ii), abiotic → biotic interaction.

At which point we encounter a phenomenon that is not currently well captured by standard evolutionary theory. It is the indirect co-evolution of the worms and the plants, via an intermediate abiotic component, the soil. As previously noted, standard evolutionary models usually "edit" out intermediate abiota, and ignore the associated abiotic dynamics, when they model the co-evolution of populations in communities. However, extended evolutionary theory need not do that. In principle, extended evolutionary theory can include any intermediate abiotic component, and its associated abiotic dynamics, in a link (iii) interaction, by measuring the ecological changes that occur in the "bridging" abiotic component as a consequence of the measured activities of an initial niche-constructing population, here the earthworms, and by then translating these changes into modified natural selection pressures for any subsequently evolving population, here the plants (Odling-Smee et al., 2003).

Even link (iv), the abiotic → abiotic link in Fig. 6.3a, could become an evolutionary link too. If an abiotic → abiotic link (iv) occurs in an ecosystem, for example in the middle of a biogeochemical cycle, and if the abiotic source in this link has previously been modified by niche-constructing organisms via a prior link (iii) biotic → abiotic interaction, then the resulting change in the abiotic "source" in the link (iv) interaction could have a "knock on" effect on its abiotic "sink" as a consequence of the activities of the initial niche-constructing population of organisms. In turn, the "knock on" effect could connect the activities of the initial niche-constructing population, to a second population, in the form of biotically modified natural selection pressures, via a subsequent link (ii) abiotic → biotic interaction. If this happened, the intermediate link (iv) abiotic → abiotic interaction would then act simply as a more complicated kind of intermediate abiotic bridge between two evolving populations. This scenario is illustrated on the right hand side of Fig. 6.3c.

6.12 EMGAs

The way in which all the links shown in Fig. 6.3c can potentially contribute to the evolution of populations in ecosystems is captured by the concept of "environmentally mediated genotypic associations", or EMGAs (Odling-Smee et al., 2003). EMGAs refer to indirect, but specific connections between distinct genotypes that are mediated by modifications to either biotic or abiotic components in ecosystems. EMGAs connect the prior expression of genetically encoded information by niche-constructing organisms, to the subsequent acquisition of "revised" genetically encoded information by any population that subsequently encounters natural selection pressures that have previously been modified by niche-constructing organisms, elsewhere in their ecosystems. EMGAs may either associate different genes in a single population. Or they may associate different genes in different populations. We think ecosystems are probably threaded through and through with EMGAs, and thus with networks of directly, and indirectly interacting genes (Van Straalen and Roelofs, 2006).

6.13 Integrating Ecology and Evolution

By now it should be clear that substituting extended evolutionary theory for standard evolutionary theory does make a difference. The comparison between Figs. 6.3a, 6.3b, and 6.3c highlights the difference. Figure 6.3a, comprises a mix of evolutionary and non-evolutionary links in ecosystems as described by standard evolutionary theory. In Fig. 6.3c, however, extended evolutionary theory describes all the links in the same ecosystem, as evolutionary. If this is correct, it means that it should no longer be necessary for ecosystem ecologists to have to choose between process-functional ecology versus population-community ecology, depending on the questions they are asking. For instance, it should not be necessary for population-community ecologists to reduce the ecosystem shown in Fig. 6.3a, to the food web shown in Fig. 6.3b, by "editing out" the abiota, before studying the evolution in the residual biota. Instead, process-functional ecology, and population-community ecology should eventually merge, as indicated by Fig. 6.3c.

In sum, the addition of niche construction and ecological inheritance to evolutionary theory allows the theory to recognise that abiotic components of ecosystems co-evolve with evolving populations, or more generally, that environments partly co-evolve with their organisms, as proposed by Lewontin (1983a). It also permits evolving organisms to act as "ecosystem engineers", by exercising at least some degree of regulatory control or influence over the energy and matter flows that occur in their ecosystems, as proposed by Jones et al. (1994, 1997).

6.14 Niche Construction and Developmental Biology

The second major area in biology I would like to reconsider, because it too has proved so difficult to integrate with standard evolutionary theory, is developmental biology, but here I can only offer some speculative suggestions. The question this time is: Would substituting extended evolutionary theory for standard evolutionary theory make it easier to integrate developmental biology with evolution?

Again, two things change when we switch from standard, to extended evolutionary theory. First, developing organisms become niche-constructing organisms capable of choosing and changing some components in their own developmental environments, primarily through phenotypic plasticity. Second, ecological inheritance is added to genetic inheritance in evolution.

Adding ecological inheritance to genetic inheritance implies that in each generation, each individual offspring organism not only inherits genes from its parents relative to its environment, but also an individual local environment that has, to an extent, previously been modified by the niche-constructing activities of its parents and other ancestors relative to its genes. This combined inheritance, genes relative to environments, and environments relative to genes, means that each offspring organism must actually inherit an initial organism-environment, [*OE*] relationship, or "niche". In standard evolutionary theory the development of offspring organisms begins with the inheritance of a "start-up" set of genes. In extended evolutionary theory, it begins with the inheritance of a "start-up" niche. The "inherited niche" is formalised in Equation 6.6.

$$Ni(t_0) = \mathrm{h}(Oi, Ei) \tag{6.6}$$

The "i" subscripts emphasise that the inherited [*OiEi*] relationship corresponds to an individual organism's developmental niche, rather than a population's evolutionary niche (Odling-Smee, 1988). Here (t_0) indicates the moment of origin of each new developmental niche, relative to each new offspring organism.

Minimally, each inherited developmental niche, or [*OiEi*] relationship, must include an initial environmental "address" in both space and time, typically influenced by parental actions, and therefore by some parental niche construction. In many species parents frequently ensure that some kind of "resource package" is also present at their offspring's initial address. For example, phytophagous insects typically choose specific host plants on which to lay their eggs, and both the eggs and the plants may subsequently serve as resources for their offspring.

So how may these changes that are introduced by extended evolutionary theory affect our understanding of individual development? Lewontin (1983a) once described the developmental process from the viewpoint of standard evolutionary theory as:

> . . ..an *unfolding* of a form, already latent in the genes, requiring only an original triggering at fertilization and an environment adequate to allow 'normal' development to continue (p. 276, his italics).

This statement was made some time ago, but in essence it is still true. Given that standard evolutionary theory only models genetic inheritance, it follows that developing organisms can only rely on the "unfolding" of inherited genetic programs, in the context of independent local environments, for the subsequent development of their phenotypes and adaptations. Developing organisms, guided by their inherited genes, adapt by responding to natural-selecting "inputs" from their local environments, but they cannot adapt by changing their local environments by their "outputs". Thus, adaptation remains a "one-way-street" process.

In extended evolutionary theory, however, developing organisms do not just respond to their environments. They also niche construct. They actively "choose", perturb and modify some components in their individual developmental environments, often to "suit" themselves. These actions introduce "feedback" into developmental processes in the same way as they do in evolution. A developing organism may modify a natural selection pressure in its environment that subsequently feeds back to affect a later stage of its own development. Waddington (1959a) captured this idea years ago with his concept of an "exploitive system" in which some of the natural selection pressures that act on developing animals are themselves consequences of the prior "choices" and perturbations of environments by those same animals.

That logically converts developmental processes from unfolding, genetically guided programs in the context of independent environments, to a process of active "niche regulation", achieved by organisms both by their responses to their environments on the basis of their inherited genes, and by the modification of their environments by their niche-constructing activities, and based on genetically afforded phenotypic plasticity. It follows, from the point of view of extended evolutionary theory, the key task for any developing organism should be to regulate its niche throughout its life, from the starting point of the $Ni(t_0) = \mathrm{h}(Oi, Ei)$ niche that it initially inherits from its parents. A developing organism should subsequently seek to regulate its niche both by responding to its environment, and by changing its environment in such ways that keep its (*Oi,Ei*) niche relationship continuously adaptive.

Ashby (1956) formulated some of the "rules" of adaptive "niche regulation" half a century ago. In particular, Ashby proposed a "law of requisite variety" which formally relates the minimal amount of variety an organism, *Oi*, must be able to deploy, relative to the variety of selection pressures it encounters in its environment, *Ei*, to protect an "essential variable", which in the case of a developing organism, means "protecting" its variable (*Oi,Ei*) niche relationship, by keeping it continuously adaptive. In practice, it probably is not possible to apply Ashby's "law of requisite variety" to entire organisms because whole organisms are too complicated. However, it might be possible to use the "law of requisite variety" to gain new insights into the development of specific sub-systems or "adaptive modules "in organisms, relative to particular sub-sets of natural selection pressures in their environments. It might be a fruitful route for developmental biologists to explore.

Thus, from the point of view of extended evolutionary theory, developmental processes are similar to evolutionary processes (Odling-Smee, 1988). Development and evolution are both "interactionist" processes. They both involve organisms responding to their selective environments, and modifying their selective environments by niche construction. An advantage of this revised approach is that it immediately assigns a major role to phenotypic plasticity in development, namely a niche-constructing role, that can connect development to evolution (Donahue, 2005). It may thereby satisfy some of the criticisms currently being levelled at evolutionary theory by developmental biologists (Jablonka and Lamb, 1995; Schlichting and Pigliucci, 1998; Gilbert, 2001, 2004; Oyama, 2001; West-Eberhard, 2003; Pigliucci and Preston, 2004; Griffiths and Gray, 2004; Miner et al., 2005). That could make it easier to integrate evolutionary and developmental biology.

6.15 Gene Networks in Development and Ecosystems

My last suggestion was inspired by Schwenk and Wagner (2004). In their recent attempt to make better sense of the relationship between developmental and evolutionary biology, Schwenk and Wagner proposed that:

> Natural selection is resolvable into "external" and "internal" components (2004, p. 394).

By external selection Schwenk and Wagner meant standard natural selection sorting between variant organisms in populations. By internal selection they meant selection pressures derived from the internal dynamics of a functioning developing organism. They contrasted the conservative nature of internal selection, with the often less conservative nature of external selection. They also claimed that the strength of internal selection is proportional to the functional integration of a system. The more integrated a system the stronger should be the stabilizing internal selection acting on its constituent characters. For Schwenk and Wagner, internal selection is a major source of the evolutionary constraints that constrain the development of individual organisms.

Their proposal raises the question whether niche construction might also be resolvable into external and internal components. The answer may be "yes". Hitherto, consideration of the evolutionary ramifications of niche construction has focused on its external component, expressed outside the body and changing an external environment, in a manner of interest to ecologists and evolutionary biologists. In principle, niche construction could also be expressed inside the body to change an internal environment and modify internal selection, in a manner of interest to developmental biologists. For illustration, in the network of regulatory interactions in the internal environment of the yeast *Saccharomyces cerevisiae*, genes encoding transcription factors interact by binding the regulatory regions of other regulatory genes, and they thereby regulate the system (Lee et al., 2002).

So is it legitimate to consider the expression of transcription factors in the internal environments of developing organisms, as a restricted form of "internal niche construction"? The only examples I know that use this idea concern the niches of stem cells (Scadden, 2006; Moore and Lemischka, 2006). They may not be typical. However, there are some intriguing hints. For example, a consequence of the putative "internal niche constructing" outputs of regulatory genes should be the maintenance, and the occasional modification of "internal selection pressures" in the internal environments of developing organisms, and therefore, the maintenance and occasional modification of developmental constraints in lineages of evolving organisms (Schwenk and Wagner, 2004).

That recalls the networks of interacting genes, based on EMGAs in ecosystems that we were considering earlier. Is there anything in common between "internal niche construction" in developing organisms, and "external niche construction" by populations in ecosystems? It could be argued that in both cases genes are expressing either simple, or complex "niche constructing phenotypes", that subsequently maintain or modify either internal or external natural selection pressures, in gene networks, that support and express complex adaptive systems. Typically, developing individual organisms are tightly regulated systems. Ecosystems and their constituent populations are usually more loosely regulated. But are there any illuminating general comparisons to be made? If there are, it will not be possible to see them until niche construction has been added to natural selection in evolution, ecology and developmental biology.

Acknowledgments I'm grateful to Kevin Laland, Anouk Barberousse and Michel Morange for criticisms of an earlier draft, and for their helpful comments and suggestions.

Chapter 7
Novelty, Plasticity and Niche Construction: The Influence of Phenotypic Variation on Evolution

Kim Sterelny

7.1 Where Does Novelty Come From? A Hypothesis

It has long been recognised that phenotypic novelty poses a prima facie problem for evolutionary theory. For there is a problem in reconciling an incremental conception of evolutionary change with the evolution of qualitatively new structures, especially complex new structures (Nitecki 1990). While it is easy to see how an existing trait could be elaborated incrementally, it is much harder to see there could be (say) an incremental, and incrementally adaptive, shift from a scaled bipedal dinosaur to a feathered and flying bird. In this paper, I shall gnaw away at this problem by weaving together three new developments in evolutionary theory. Those developments, in combination, do not completely resolve the problem of novelty, but they make it more tractable.

The argument begins by rehearsing an "evo-devo" argument: the mechanisms that generate variation cannot be black-boxed, for unless those mechanisms are isotropic, generating variation densely and evenly around a lineage's existing location in phenotype space, the mechanisms that generate variation will influence the trajectory of evolution (Raff 1996; Arthur 2004). And while there are many open questions about variation, there is no a priori expectation that variation is typically isotropic. I then link this now familiar line of argument to an emerging emphasis on the role of phenotypic plasticity in evolution. In different ways, Mary-Jane West-Eberhard and Marc Kirschner and John Gerhart have argued that plasticity is central to evolvability (Gerhart and Kirschner 1997; West-Eberhard 2003; Kirschner and Gerhart 2005).

Their idea is this. Phenotypic plasticity pre-adapts lineages to evolutionary change, by connecting the development of distinct organ systems. If organisms are phenotypically plastic, then the functional integration of different systems—for example, those that determine bone growth in limbs and those that lay down the circulatory system—must be sensitive to the contingencies of their development. Limb development requires simultaneous and co-ordinated development in other organs

K. Sterelny (✉)
Professor, Philosophy Programme, RSSS, Australian National University, Canberra, Australia
e-mail: kimbo@postbox.anu.edu.au

A. Barberousse et al. (eds.), *Mapping the Future of Biology*, Boston Studies in the Philosophy of Science 266, DOI 10.1007/978-1-4020-9636-5_7,

and tissue systems: cartilage, muscle tissue and attachment points, innervation of soft tissues; circulatory connections to tissues and bone barrow. If bone structure or muscle mass is plastic, responding to signals from the environment, co-ordinated systems must be plastic too, responding to signals from the systems developing with them. For the organism does not "know in advance" how much muscle tissue to build, or how densely the capillaries should be packed into the tissues that surround the bone. Co-ordination must be managed via cross-talk, as these systems develop. They must be able to fine-tune their own development in response to signals from the systems to which they attach. This same sensitivity of integration to the contingences of development will make functional integration possible in the face of genetically-caused changes in crucial organ systems. The systems that build muscle tissues and attach those tissues to the limbs will do so appropriately in the face of genetically modified changes in bone size or shape. Hence plasticity pre-adapts for evolvability.

In adaptable organisms, environmental change generates novel phenotypes: organisms respond behaviourally, physiologically and morphologically to new stresses and opportunities. Novelty appears by *environmental induction*. Such novelties have no effects on the germline are not inherited. But they can still have evolutionary effects. Once those novel phenotypes appear, they are subject to a selective sieve. Some will disappear. But if the environmental inducer continues to be present, some of these novelties will reappear in the next and subsequent generations, rebuilt by the same mechanisms of phenotypic response. One consequence will be genetic accommodation in its various forms. The novel trait itself will be fine-tuned in various ways by genetic change. Likewise, other aspects of the phenotype will be adaptively adjusted to the novelty. So evolutionary change (on this view) often involves a double pulse: new phenotypes often appear first as a result of phenotypic response to environmental novelty. They are then followed by genetic accommodation.

The final idea is to link phenotypic plasticity to the source of environmental change. As we have just seen, phenotypic plasticity is an evolutionarily consequential response to environmental change. Such environmental change has typically been seen as an outside influence on a lineage. (Eldredge 2003) is a typical recent example of this perspective on change and its evolutionary consequences. Evolutionary change is essentially a response to environmental changes impinging on a system from the outside. The extent of evolutionary change depends on the intensity and geographical extent of physical disturbance. This picture of change and where it comes from underplays the active role of organisms and populations in constructing their own environment. Richard Lewontin was the first to emphasise the importance of organisms' agency in selecting and modifying their environment; recently, niche construction theorists have developed and extended Lewontin's insights (Lewontin 1982, 1985; Odling-Smee et al. 1996; Jones et al. 1997; Odling-Smee et al. 2003). We now get the link between niche construction and the evolution of novelty. For though phenotypic plasticity is sometimes a response to extrinsic changes in the environment, it can also be a response to changes the lineage itself has caused. Niche construction—the effects organisms have on their own world—is important for the ways it modifies selection pressures, and hence the ways existing variation is sifted (Odling-Smee et al. 2003). But it is also important in generating the variation on which the evolution of novelty depends.

7.2 Biased Variation and Evolution

Rudy Raff's *The Shape of Life* signalled a major reorientation in evolutionary biology: the integration of developmental biology within contemporary Darwinism (Raff 1996). Developmental biology is relevant to evolutionary biology because developmental biology is our theory of phenotype possibilities: of the potential variation that is available to selection. Developmental biology is in the business of identifying the $G \Rightarrow P$ map, and its potential permutations. It took some time for evolutionary biology to accept that it was crucial to understand the $G \Rightarrow P$ map, because a crucial assumption was made about its nature. The working assumption of the modern synthesis was that variation in natural populations was typically isotropic: the variation in a trait is distributed densely and without bias around its current mean. If this assumption were right, while the mechanisms through which a genome generates a phenotype would be interesting in their own right, evolutionary biologists could reasonably idealise away from these complications.

It is not likely, though, that this idealisation is appropriate: it is tied to Fisherian models which supposed that the development of a phenotype trait (and hence variation in that trait) depends on large number of small effect genes. But mutation is not restricted to point mutations: mutation which might substitute one amino acid for another. Mutation can result in movement, duplication, inversion and deletion of DNA sequences, and hence can result in changes to gene regulation and to shifts in reading frames as well as changes to the amino acids that are transcribed from the new DNA. Mutations which make a significant difference to continuously varying traits—"mesomutations"—are likely to be important in many evolutionary changes.[1] These may not be distributed normally around current phenotype values.

Such a bias in its supply is potentially relevant to evolutionary trajectories.[2] Wallace Arthur has driven this point home with exceptional clarity. As he notes, it is likely that the supply of variation is structured, and that structure matters. Consider a diagram plotting mammal species on a two-dimensional grid. The y-axis gives the

---

[1] For a good discussion of these issues about mutation and their significance to our view of evolution, see Leroi, A. M. (2000). "The Scale Independence of Evolution." *Evolution and Development* **2**(2): 67–77.

[2] So too is the spread of variation. Imagine a rodent population with a mean mass of 500 grams. Around this mean, the fitness surface is flat. Within a certain bound, mass makes no difference to fitness. But in the vicinity, there are two higher local optima: 600 gram rats are fitter than 450 gram rats which in turn are fitter than those of 500 grams. The evolutionary trajectory of this population depends not just on bias in the generation of variation around the current mean, but also its spread. If the mechanisms of generation generate very little variation, the population will stay where it is. If it generates variants within 50 but not 100 grams of the current mean, the population will migrate to the lower local optima. If the spread is great enough for variants to intersect with the slopes of the higher peak, the population will find that peak.

length of front-limbs; the x axis the length of the back legs. Since most species have front and back legs of near-equal length, the result is a pattern where most mammals fall close to the diagonal from bottom left to top right (Arthur 2004, p. 99). Why are mammal phenotypes clustered in this way? The supply of variation to selection is likely to be biased, for many developmental processes affect both sets of limbs. Selection might play some role: perhaps few life-ways reward elongated front limbs. Notice, though, that it is much harder to give a plausible selective explanation of why mammals mostly have the same number of fingers and toes. So while it is not impossible to alter front to back ratios—kangaroos have much longer back legs than front ones—that ratio does not change easily. Adaptive complexes like kangaroos are difficult to reach, and difficult to reach because of structure in the supply of variation. The natural supply of variation will not provide much variance in length between the two limb pairs, and hence adaptive peaks with unequal lengths will be too far away from current variation pools for them to be available. Likewise, while it is possible to lose toes from hind legs, it is difficult to adaptively optimise finger and toe numbers to their different roles, even when the functional demands on front and hind digits are quite different.[3]

So bias in standing variation can influence evolutionary trajectory, by making apparently nearby optima inaccessible. Similar considerations apply to the supply of new variation to selection. Suppose a population finds itself on a flat fitness landscape near incompatible local optima. Such a population is "waiting for a mutation". Different possible mutations will extend standing variation to the foothills of one of those local optima, after which selection-driven hill climbing will do the rest. The trajectory of the population will depend on which variation arrives first, and one might be more probable than another, even if the two potential variations are an equal distance in phenotype space from current phenotypes. Phenotype distance may not correspond well with genotype distance. Moreover, some phenotypes are multiply realisable, as the phenomenon of phenocopies shows: different genomes can give rise to the same phenotypic outcome.[4] Other phenotypes have a single genetic profile; their development depends on a specific set of genes. Phenotypic variants that can be reached by a number of tracks through gene space are more likely to appear than those that depend on a specific track.

So the first crucial point in the argument of this paper is that if the supply of variation to selection is biased, that supply can play a pivotal role in determining evolutionary trajectories. It does so if variants at points in phenotype space that are equi-distant from current phenotypes are not equi-probable.

---

[3] Though the genetic details of digit evolution and development are quite complex, with influences from multiple Hox genes: for a survey of some of these complexities, see Zakany and Duboule (1999).

[4] This phenomenon may be quite extensive. West-Eberhard mentions that upwards of 50% of experimentally induced gross chromosomal rearrangements in *Drosophila* have no phenotypic effect. While some of those no-effect results will be the result of changing junk DNA, many are inert because changes are buffered by canalising mechanisms (p. 19).

7.3 Phenotypic Plasticity and its Significance

We have just seen that the supply of variation has the potential to play a driving role in evolution. I now link this idea to phenotypic plasticity. An organism is phenotypically plastic if its genome maps onto different phenotypes in different environments. It is adaptively plastic if these differences (over some environments) enhance its fitness. Human physiological adaptation to life at high altitudes is an example of adaptive plasticity. Given this notion of plasticity, this section has two aims. The first is to explain the connection between plasticity and evolvability. The second is to show that the mechanisms of phenotypic plasticity play a crucial and biasing role in supplying that variation: there are indeed biases in that supply. In part, that is because environments play a crucial role in inducing variation, not just selecting it. This idea and its evolutionary consequences have been explored by Mary-Jane West-Eberhard and, in collaboration, Marc Kirschner with John Gerhart (Gerhart and Kirschner 1997; West-Eberhard 2003; Kirschner and Gerhart 2005). West-Eberhard, Kirschner and Gerhard all argue that phenotypic plasticity has profound consequences for evolution. Within-generation plasticity enhances evolutionary plasticity. Lineages are evolutionarily plastic because organisms are phenotypically plastic.[5]

It is no accident that organisms are plastic; plasticity reflects the complexity of the causal pathway from genetic material to expressed phenotype, and the many points at which contextual factors can affect these pathways, and hence plasticity is, as Schlichting notes, the initial condition.[6] This initial condition can be selectively modified, sometimes to reduce sensitivity to environmental factors, sometimes to channel it. For organisms (especially multi-celled organisms) can be advantaged by plasticity. (i) They develop, and in many lineages development involves profound morphological change: thus core organ systems must continue to function and co-ordinate with other systems through changes in those other systems. (ii) Though tolerances vary, all organisms must be able to recover from damage. Organisms must be organised in ways that buffer rather transmit the effects of local failure and damage. (iii) Individual organisms do not live in constant environments: some organisms are mobile, and even immobile organisms experience environmental variability as their local circumstances change daily and seasonally. Yet different environments require different physiologies. (iv) Organisms live in unpredictable environments: plasticity extends the range of environmental stresses to which organisms can accommodate. (v) Organisms develop, and development does not yield

---

[5] For the purposes of this paper, I am neutral on whether phenotypic plasticity has been selected to enhance evolutionary plasticity. Richard Dawkins has suggested that there may be some form of lineage-level selection in favour of evolvability: Dawkins, R. (2004). *The Ancestor's Tale: A Pilgrimage to the Dawn of Life*. London, Weidenfeld and Nicholson. While sympathetic to that idea, I make no commitment to that view in this paper.

[6] Thus Schlichting remarks that even nucleic acids and enzymes have reaction norms: they will function only within a restricted band of temperature and pH, a band which can itself be selectively modified, as the production of so-called "heat shock" proteins by Antarctic, cold-adapted fish at 5C strikingly show: Schlichting, C. (2003). "Origins of Differentiation via Phenotypic Plasticity." *Evolution and Development* **5**(1): 98–105, p. 99.

phenotypic uniformity because development resources both within and without the embryo are not uniform. Developing embryos will be exposed to differing environmental fluxes, and be supplied with differing nutrient packages. Nor is the genetic environment predictable: elephant seal heart-building genes need to initiate the development of a functioning heart in both females and in the enormously larger males. Moreover, once a certain level of complexity is reached, development cannot be precisely controlled. Thus a given component needs to be able to work in somewhat different internal environments. Its systems of signalling, co-ordination and linkage must be able to cope with somewhat varying organizations of internal components.

Plasticity is mediated by a combination of modularity and "exploratory behaviour" (as Kirschner and Gerhart label it). Developmental modularity is crucial: unless the development of organ systems are decoupled from one another, a disruption anywhere will ramify through the system. Even so, organ systems must be appropriately connected to one another if the organism is to function. For example, the vascular system of mammals is extraordinarily complex: cells are never more than a few cell diameters away from a capillary. Yet the precise plumbing cannot be pre-specified: it must be sensitive to bone and muscle growth. So its development is patterned, but responsive to developmental signals from elsewhere in the organism. Thus capillaries are provided by a process of over-supply and selective attrition. Wherever muscle development is dense and hence the need for oxygen flow is great, less of the over-supply will be deleted.

West-Eberhard has some wonderful examples of phenotypic adjustment of this kind. Her flagship example is a two-legged goat: a goat that was born with dysfunctional front legs and which acquired a kangaroo-like gait. The important point about this example was the accommodation of the rest of the phenotype to this reorganisation of locomotion: there was an extraordinary suite of muscle, nerve and bone reorganisations to accommodate the different stresses, loads and movements (West-Eberhard 2003, pp. 52–53). This is a very extreme example, and no doubt the goat would never have lived had it not been born into a domestic environment. But West-Eberhard gives a series of other examples. In human populations there are many pathologically developed hearts, with arteries, veins and valves in non-standard places. While these may not be optimal, these developmental pathologies are not instantly fatal. The rest of the phenotype accommodates to them, connecting the system functionally to the circulatory and respiratory system. In development, bone and skin accommodate to the loads to which these systems are subjected (see e.g. West-Eberhard 2003, p. 35), and the friction they endure. Bone is hard at a time, but it in its growth and form it is extraordinarily plastic.

There are crucial cellular processes which manifest similar adaptive plasticity: for example, the mechanism in mitosis that ensures that each daughter cell receives the right chromosome complement ("spindle formation") is adaptively plastic. It is not and cannot be pre-programmed with information specifying the location in the dividing mother cell of these chromosomes. So the microtubes that usher them to the daughter cells explore from the centriole, and are stabilised if they connect with a chromosome. Thus mutations which increase chromosome number need not be fatal: the mechanisms of mitosis can accommodate to this change without needing a further genetic change. The phenotype adjusts to the unpredictable at all organization levels of the organism, through cells (cytoskeleton and microtubules); vascular system; to the behaviour of the organism as a whole (as in learning).

The existence of these mechanisms allows phenotypic adjustment to changes elsewhere in the phenotype without correlated genetic change. Thus if selection for size or strength results in genetic changes which increase the neck muscle mass of a male deer (via, say, sexual selection), there is no need for further genetic changes to ensure that those muscles are adequately serviced by the circulatory system of the animal. These mechanisms of phenotypic accommodation remove a crucial roadblock that would constrain the evolution of novelty. Limb evolution, for example, requires simultaneous changes in many tissues: bone and cartilage; muscle position and perhaps mass relative to bone; muscle attachment points; innervation of new muscle tissue; vascular change. Without exploratory mechanisms, the evolutionary co-ordination problem would severely constrain adaptive change. This problem of correlated change was one reason for thinking evolutionary change must be gradual. A large-effect mutation increasing neck muscle mass on a male deer would decrease fitness unless there were correlated increases in bone mass and shape, and in the supply of blood to those tissues. A small increase in muscle mass might still attach successfully in an unchanged context, while selecting for "catch-up" changes in the associated systems. Thus the deer lineage might inch towards a new male phenotype. But if, within limits, a whole suite of systems are developmentally responsive to signals from their partners, a large effect mutation that (say) increased horn size might be favoured. For its development would trigger the appropriate adjustments of neck muscle, bone and circulatory systems. Phenotypic accommodation finesses the problem of correlated change: a genetically-caused modification in one system need not wait for a genetically-caused change in associated systems, even when both must change for either to be adaptive.

So, first, mechanisms of phenotypic plasticity enhance evolvability by enabling phenotypic adjustment to genetically-caused changes in an organism. These mechanisms act as *change amplifiers*: small genetic changes (changes that directly affect only one component of an organism) can result in a suite of adaptively correlated changes. Thus a small change in G-space can map onto a large change in P-space. Consider, for example the gracile human face: our teeth, face and jaw are less robust than those of earlier hominins and our closest great ape relatives. We do not know the genetic distance between contemporary humans and early hominins, but we do know that the genetic distance between the chimps and modern humans is surprisingly small, given the impressive phenotypic differences between us. These facts about the $G \Rightarrow P$ map point to a possible explanation. The evolutionary restructuring of our jaw, face and teeth might depend on only a few genetic changes which affect directly only a few elements of the facial complex, together with correlated change mediated by the mechanisms of adaptive plasticity. This is no mere conceptual possibility: it is known that bone growth is very sensitive to the loads imposed on bone in development. It is phenotypically labile. Indeed, there is some suggestion that a common contemporary human dental problem may be due to an insufficient load on our lower jaw in development, due to a shift in childhood diets to softer foods (Gilbert 2001, p. 8).

Plasticity enhances the flow of variation to selection, but it also biases that flow. Some aspects of phenotypes are easily altered; others are not. The switches that control developmental subsystems—that turn them on, and connect them one to another—are easily alterable. In vertebrates, sex determination can depend on a variety of environmental or genetic switches. As lineages evolve, control can be passed from genes to environment or vice versa. But once initiated, the developmental mechanism is highly conserved. So while in some ways the mechanisms of sex regulation in vertebrates is very labile, in other respects it is conservative. The underlying molecular mechanism is conserved, depending on an estrogen circuit with two stable states and the SF-1 protein trigger. There is much lability in how that trigger is pulled, but once pulled, the cascade has a conserved structure across the vertebrates (Kirschner and Gerhart 2005, pp. 90–96). In general, while the regulation of developmental modules varies, the modules themselves are conserved. Each phylum seems to have a common and distinctive pattern of compartmentalisation in the early embryo; a pattern which sets up its stock of developmental modules (Gerhart and Kirschner 1997, Chapter 7). So lineages are at once more and less variable than the isotropic model predicts. Small genetic changes can generate a suite of correlated changes: more change than the isotropic model expects. But some crucial aspects of the phenotype are almost invariant.

The mechanisms of phenotypic plasticity thus modify and in some ways accentuate the variability available to selection. But as Mary-Jane West-Eberhard argues at (great) length, they also make the environment itself a direct source of evolutionarily significant variation. The environment induces variation; it does not just select genetically-induced variation. For one thing, environmental change can make both new signals and new resources available to developing organisms. For example, many insects protect themselves with poisonous chemicals without manufacturing those chemicals: they are found in the plants they eat, and the insects have evolved the capacity to tolerate and sequester them (West-Eberhard 2003, pp. 501–505). It is likely that such novelties often have environmental causes. The environment changed as plants evolved chemical defences. But some insects in the environment managed to accommodate ontogenetically to this change, perhaps by storing these chemicals relatively harmlessly in their tissues, thus leading to a subpopulation with a novel phenotype. Subsequently, this capacity has been elaborated and fine-tuned by genetic accommodation, giving us the array of chemically protected butterflies and other insects that we now find.[7]

[7] Of course, this is not the only possible trajectory. There is an alternative possibility: the plant develops a toxin which drives the ancestral caterpillars from it: they retreat to alternative food sources, to wait on a favourable mutation, conferring metabolic protection against the toxin. Once this mutation arrives, eventually, the caterpillars can reinvade, and once they have done so, natural selection on continuing genetic variation elaborates this adaptive complex. In this scenario, the novelty is selected by the environment but not induced by it. The toxin-protected leaves are part of the selective environment, but no part of the explanation of the origin of the novelty. Generalists might evolve toxin resistance by this route but specialists are less likely to do so. For the mutation-lead event has to be combined with the fortuitous accident of host-switching for the mutation to be selectively favoured.

It is true, of course, that environmentally-induced variations of this kind do not thereby rewire the organism's genome, thereby transmitting the novelty to the next generation. Even so, environmentally-induced variation is often evolutionarily important. One important contrast between mutation-driven and environmentally-induced novelties concerns the size of the initial population of organisms with the new phenotype (novels). Initially a mutation-triggered novelty will be very rare; carried only by the organisms in which the mutation occurred and some of its direct descendants. An environmentally-induced novelty is likely to characterise a significant sub-population right from the beginning: since its development depends only on widely-shared mechanisms of phenotypic plasticity, it is likely to develop in many of the organisms exposed to the environmental trigger. The novelty is therefore much less likely to drift out of existence. Its effects on fitness are scrutinized much more effectively, not just because the novel phenotype is less likely to drift to extinction, but also because there is likely to be variation in the novel phenotype itself. For those with the novel phenotype are likely to differ both in the details of the new trait and in how well the rest of the phenotype accommodates to the novelty.

Moreover, some mechanisms by which phenotypes adjust to new environments are biased in the direction of adaptive response, and thus environmentally induced change is an especially important route to the evolution of adaptation. For in many cases the underlying mechanism is some version of trial and error, with the selective stabilisation of success. The proximate signal of successful learning—physiological reward—is not, of course, perfectly correlated with fitness-enhancing actions. But it is certainly to some extent correlated with fitness, and the same is true of other exploratory mechanisms. For while learning is an obvious example of adaptive accommodation, it is by no means the only one, as West-Eberhard's wonderful example of the two-legged goat illustrates. Finally and crucially, the environmental changes which induce a novel phenotype in generation N—say, the invasion of a plant species with new toxins in its leaves—continue to act. So the novel phenotype will be induced again in the next and subsequent generations. The novelty is induced as a multi-generational and varying subpopulation. The novel phenotype is generated in significant numbers, over several generations, and it has fitness consequences. So it will trigger processes of *genetic accommodation*: there will be selection on genes which modify the novel phenotype. It is this accommodation—the second pulse of change—which transforms the environmental induction of a novel phenotype from a merely ecological event to an evolutionary one.

Genetic accommodation can take many forms. Suppose, for example, that the novel phenotype is deleterious: the caterpillars that eat the new plant with its new toxins survive and develop, but have reduced fertility as butterflies. Then genes which modify or suppress those effects will be selected. Suppose that the phenotypic effects are on balance positive but with negative side-effects; the toxins protect the caterpillars but at some cost to fertility. Then there will be selection for genes which suppress the side-effect. Suppose the new phenotype is superior: the caterpillars can store the toxins harmlessly, buying protection without cost. Then genes that make the development of this phenotype more likely will be selected: for example, genes which make the plant more perceptually salient to the caterpillar.

The much-discussed Baldwin Effect is a special case of genetic accommodation to an environmentally induced phenotypic change: accommodation that makes development of the new phenotype less contingent on specific signals from the environment. Waddington's examples of genetic assimilation, likewise, are a special case of genetic accommodation that makes the development of a new phenotype less sensitive to specific features of the environment (Waddington 1959b).

Avital and Jablonka have sketched a further special case: their "assimilate and stretch" model of the evolution of behavioural complexity. Their model applies to those agents that face cognitive bottlenecks through being under selection for the elaboration of cognitively demanding skills. Consider such traits as complex bird song, bowerbird bower building, language. Suppose that a species of bowerbird is under selection for bower-quality: for an individual male bird, the more elaborate the bower, the better. And suppose that learning to build a bower is cognitively demanding. A bird's cognitive resources constrain its capacity to build an elaborate and attractive bower. If the basic bower building skills no longer had to be learned, the cognitive bottleneck on such birds would be eased, and their cognitive resources would then stretch to a more complex bower. Bower-building as a whole would have both innate and learned elements, even as bowers became more elaborate. Further canalisation would then allow further stretching to more elaborate bowers yet, though still built through an ensemble of learned and innate elements. And so on. At each stage the baseline is raised. But female preference for elaboration selects for further canalisation, and hence we get increasingly elaborate versions of the trait in question (Avital and Jablonka 2000; Jablonka and Lamb 2005).

Avital and Jablonka took their mechanism to describe one possible, though important, dynamic between canalisation and learning. But our hypothetical (though not very hypothetical) butterflies suggest that it might have more general application. It applies to any circumstances in which (a) there is enduring selection on a trait along a dimension. For our caterpillars, the more toxin-resistant they are, the better. They are better protected and have more access to food. (b) Though there are mechanisms of phenotypic plasticity that shift trait value in the right direction, there are resource limits on these mechanisms. (c) There are genetic changes which would shift the norm of reaction in the right direction, easing those resource limits. For example, there might be genetic changes which increase the number of cells in which the toxin can be stored; which up-regulate the production of antidotes, or which increase the extent to which vulnerable eggs are buffered against the toxin's effects. Toxin-resistance would still depend on mechanisms of phenotypic accommodation, and the toxin would still be an environmental stress. But the overall effect of genetic change would be to increase the quantities of toxin the caterpillars could safely ingest and store. It is possible that the "assimilate and stretch" model might have general application in explaining directional changes in a trait over time.

Avital and Jablonka's work on learning and behavioural traditions leads naturally to the next theme of this paper: the role of the organism itself in causing the environmental changes which in turn cause changes in the supply of variation to selection. For they point out that the upstream generation profoundly affects the learning environment of the downstream generation, and hence profoundly affects behavioural competence, and variation in behavioural competence, of that downstream variation. The role of parents in shaping the behaviour of their offspring is obvious in species like ours, where both parents and children have psychological capacities that have been selected to enrich and to stabilise the flow of information across generations. But those capacities are rare: they characterise few species. It is important to notice that stabilised behavioural traditions are not restricted to those species that are highly adapted to cultural learning. The ordinary ecological activities of parents structure the learning environment of juveniles. When offspring live with their parents, as they do in many species, adult activity structures the learning environment of the juveniles. Even in those species without special adaptations for high-fidelity cultural transmission, the result can be the reliable flow of information across generations. Once chimpanzees (for example) begin to termite fish regularly—once the harvesting termites becomes a regular aspect of their foraging—juveniles will have many opportunities to acquire termite fishing skills by undirected trial and error learning (Avital and Jablonka 2000). The initial innovation might have been a low probability event, but once it is made, it changes adult lifeways, enhancing the learning opportunities for the next generation. Innovations can thus be entrenched in local populations. Behavioural traditions of this kind are important expansions of adaptability, because they prepare generation N+1 phenotypes for local environmental conditions: conditions too local in space and time to be tracked by specific, genetically mediated adaptations.

The learning environments of mixed generation species are not found; to a very considerable extent they are made. The experience of juveniles depends profoundly on adult activity: their learning world is a made world. As we shall see, this is true not just of juvenile learning in populations with a mixed age structure. The world organisms experience is in part their own product.

7.4 The Organism-Environment Developmental Loop

Genetic accommodation in its various forms initiates cascades of gene changes which in turn result in further phenotype changes, both directly and through further phenotypic adjustment. It thus makes environmentally-induced phenotype novelties evolutionarily significant. Sometimes the environment changes for reasons independent of the local biota, and that biota responds. The climate or the sea level changes; new species migrate in; others become locally extinct. But organisms often change their own environment. Sometimes these changes are mere side-effects of their ecological and reproductive activity: niche construction in the broad sense. But organisms often do not just change their environment, they enhance their own fitness by adapting it (for the distinction, see [Sterelny 2005]). Thus termites, earthworms, parasites, trees do not live in the world as they find it. These organisms and others in part remake their world. As the theorists of niche construction have shown in such compelling detail, in acting on their environments, these organisms profoundly change the selective forces acting on them (Odling-Smee et al. 1996, 2003). Termite mound and tunnel systems take most insectivores out of the termite danger equation, and moderate the ranges of temperature and humidity to which they are subject. The environmental changes termite engineering initiated have selected for modification in termite phenotypes until virtually nothing is unaffected by life in the mound (Turner 2000). There is, then, a selection-mediated feedback loop from the organisms in a lineage to further evolutionary change in that lineage. If the argument of this paper is right, there is an additional feedback loop. The agents that compose a lineage can influence *the supply of variation* to selection and hence (as we have seen) its evolutionary trajectory.

Consider first *niche choice*. Organisms often choose their environments: they recognise and respond to environmental stimuli, and those responses result in sampling their physical surrounds selectively. One consequence is that they sometimes find themselves colonising new adaptive zones, both ecological and geographic. When organisms disperse to new habitats—for example, birds, insects or seeds being blown to islands—they clearly experience new selective environments. But such dispersing organisms also experience new physiological environments and (if the initial colonisation survives) their offspring will experience new developmental environments. Often the same will be true when organisms colonise new ecospace: for example, when phytophagous insects switch hosts. Exploring ecospace has been very important. Such exploratory behaviour has been the leading edge of world-transforming change: for example, the invasion of sea-floor substrates at the beginning of the Cambrian and the colonisation of terrestrial habitats. Major shifts of adaptive zone result in major morphological and physiological re-organization, but they are almost certainly initiated by dispersal and exploration, and first made possible by mechanisms of phenotypic adjustment to altered environmental conditions.[8] The exploration of new ecospace, and especially involuntary exploration due to forced dispersal through storms and the like, taking organisms to quite unfamiliar environments, is likely to generate in the survivors new variation. Genotypes which are equivalent in one environment need not be equivalent outside the normal range. Stress can result in novel phenotypes, making previously unexplored ranges of norms of reaction relevant to selection and exposing once-hidden genetic variation to selection.

West-Eberhard argues that the importance of these phenomena has been understated by lab-based genetics. By narrowing the class of rearing environments, genetic control looks simpler than it really is. As the class of rearing environments becomes broader as organisms spread through a variable habitat, multiple influences on phenotype become more common. This facilitates genetic accommodation to environmentally-induced novelties, for there will typically be many genes influencing the form, extent, timing and phenotypic adjustment to a novelty

[8] For suppose the lucky mutation (or mutations) that increased tolerance to (say) temporary exposure on mud surfaces and the like came first. There would be no selection for these mutations unless they were accompanied by a behavioural change to take advantage of these increased tolerances.

(West-Eberhard 2003, pp. 505–515). So niche construction promotes variability by exposing organisms to stressful environments, and, more generally, environments outside the range of previously evolved canalisation mechanisms. Those are also populations subject to strong selection pressures. As selection intensifies, so too variability in the population is likely to increase.

Niche construction in the broadest sense is probably most important because of its role in exposing more of the potential norm of reaction to selection. Niche engineering—niche construction in a narrow sense—plays a different role, for it controls and homogenises the developmental environment, as I will now argue. For consider *the norm of reaction concertina.* The effects on phenotypes of novel environments are a special, though extreme, case of a general pattern. The idea of a G $\Rightarrow$ P map is misleading. For the phenotypic effects of genes are context-sensitive. As we vary the environment, the same set of genes map onto different phenotypes. For example, amongst the insects there are many seasonal polyphenisms: the same butterfly can look very different, according to its hatching season. (For a brief overview, see Gilbert 2001). The sensitivity of gene effects on phenotype is often captured through the concept of a reaction norm (see, especially, Schlichting and Pigliucci 1998; Lewontin 2000a). We represent a reaction norm when we graph the phenotypic effect of gene/environment combinations. Occasionally reaction norms are flat (showing the phenotypic effects of a gene are insensitive to environmental variation) but that is the exception rather than the rule. The environmental heterogeneity that a population experiences is relevant to the supply of variation it presents to selection. If a population experiences a homogeneous environment, it will expose less variation to selection; if it experiences a heterogeneous environment, it will expose more variation to selection. In turn, that heterogeneity depends in part on both niche choice (including accidental choice) and niche engineering by the agents that compose that population. Both heterogeneity and homogeneity can be *made*, not merely experienced.

Niche construction helps determine the relevant range of the reaction norm: which areas of E intersect with G to produce P. Niche choice tends to extend the relevant range of norms of reaction: a given gene experiences more environments as a consequence of active exploration and choice. Phenotypic plasticity accentuates this exposure of the reaction norm; it expands the concertina. Plasticity in generation N tends to expose more of the norm of reaction to selection at generation N+1, because plastic organisms have a broader range of environmental tolerance. They reproduce in this broader class of environments and that widens the class of Es in which different Gs act.

Ecological engineering—the most dramatic and obvious form of niche construction—tends to narrow the relevant range, though at the same time that narrowed range is shifted by comparison to the ancestral state. Burrows, warrens, beaver lodges and similar structures buffer the effects of environmental variation: they make the *experienced environment* of those living within more homogeneous than they would otherwise be. Think again of termites. One consequence of the evolution of eusociality is that the social world of the nest and the physical environment of the nest have become crucial and stable features of the termite world. That world is very different from the world of termite ancestors: so there has been a shift in the range of E in which termite eggs and larva develop. But for a given termite species, the nest environment is stable. There is some variation in size and construction from nest to nest. But many species have distinctively shaped nests with a characteristic architecture, so termite genes are expressed in a narrowed range of environments.

Sometimes both the extension and the channelling of reaction norms can be combined in a single process, as in Schlichting's model of the origins of cellular differentiation in multicellularity. Imagine a clone of cells aggregating after mitosis rather than drifting part, perhaps initially to avoid being engulfed by other protists. The central cells in the cell ball will have a uniform and narrowed environment, consisting just of others like themselves. But edge cells will experience the world differently: with an internal interface as well as one with the less predictable and more hostile external environment.[9] The two cell populations will, therefore, explore different regions of their common reaction norm and may not be phenotypically identical. Differentiation may have begun as a by-product of differential exploration of a common norm of reaction, before its selective elaboration (Schlichting 2003).

Novelty and Squeezing The Concertina. This suppression of environmental variation may well be important to the evolution of developmental control over complex phenotypes. Maynard Smith and Szathmary emphasize the fact that evolutionary transitions are transitions in the kind and quantity of information transmitted across the generations (Maynard Smith and Szathmáry 1995). For phenotype complexity depends on a cross-generation flow of information and phenotypic complexity. If the complex phenotype is heritable, the cross-generational signal has to be both rich and of high fidelity to control a complex developmental process (Ridley 2000). But since genetic signals can do nothing more than initiate developmental cascades, genetic signals must be inserted into a sufficiently structured and predictable developmental environment. So in the typical case, an organism of generation N influences the phenotype of its offspring in N+1 in two ways. It contributes to the genetic complement of the N+1 individual. And it engineers the developmental environment of those genes; it structures the way those genes are read. The egg (and its equivalent with plants and fungi) is a structured system: adapted both to function in an environment and to provide an initial set of triggers for gene expression. Eggs vary remarkably from one another; indeed, perhaps they vary more than the early embryos into which they develop (Gerhart and Kirschner 1997, Chapter 8, Raff 1996, Chapter 6). But for all their differences, animal eggs must provide positional information that controls the initial differentiation of the early embryo. So eggs vary, but all are maternally engineered (for a review, see Gilbert 2003c, Chapters 7, 8). Some parental engineering begins and ends with the egg, but many organisms engineer developmental environments more extensively. This structuring

[9] The same is true, though to a lesser extent, if we think of differentiation as evolving in a cell film, so that each cell has feeding access to the external environment: central cells in the film will have a more protected, more cell-neighbour environment than those on the edges of the film.

takes many different forms. Some parents provide key developmental resources: for example, many invertebrates have elaborate systems for the transmission of symbiotic micro-organisms from offspring to descendant. Development may be buffered by specifically adapted internal or external structures: cocoons, wombs, nests, burrows are engineered developmental environments. Even when the N-organism does not engineer the developmental environment of the N+1 organism, they quite often select them: many female Lepidoptera, for example, choose their egg locations very carefully.

Developmental engineering plays a crucial role in the evolution of novelty. On one view of the Cambrian evolution of animal life, the pulse of differentiation depended on the transformation of animal developmental biology. On this view of the Cambrian, the earliest bilaterian animals had a simple developmental system in which the cells of the developing embryo differentiated immediately, and divided and moved to their adult positions under local control. But while developmental programs of this power sufficed for the reliable development of a very small, not very complex worm-like animal (somewhat akin, the idea goes, to the planktonic larval forms of some living animals) it does not suffice for larger and more complex animals: animals with appendages and complex sensory systems. The evolution of more complex animal life depended on the evolution of populations of "set-aside cells"—cells whose cell fate is not determined at the point at which development begins. And it required the evolution of compartmentalized development; a developmental organization in which the developing embryo divides into a series of regions, which then develop semi-autonomously (Peterson and Davidson 2000; Davidson 2001; Erwin and Davidson 2002). If these defenders of the "developmental revolution" hypothesis are right, it follows that the adapted egg played a central role in the Cambrian explosion and thus in the invention of complex bilaterian animal life. For this elaborated developmental recipe depends on positional information in the egg—information from which the initial embryo geography of autonomous compartments can be derived. Bilaterian complexity, on this view, requires downstream developmental engineering of the egg.

The developmental revolution hypothesis remains controversial. But Brett Calcott has pointed to more general considerations with a similar theme (Calcott 2006). Complexity depends on reconciling two competing demands of differentiation and shared evolutionary identity. The generation of benefit requires differentiation; the division of benefit requires identity. It pays cells to stick together because of collective synergies in survival and in gathering resources. But this phenotypic power of complex animal and plant life depends on specialisation and the division of labour, and hence on cellular differentiation. August Weismann, the great German evolutionary theorist, supposed that cellular differentiation depended on the differential distribution of heritable factors from the germ-line: nerve cells are different from skin cells because they contain different genes, not because they have different genes active (Maynard-Smith 1989). In retrospect, perhaps it is no accident that Weismann's hunch was wrong, for a genetically chimerical body might well be an unstable alliance. Differentiation with identity, though, depends on signals from the previous generation. We see this in eusocial insects: the difference between queen and worker is a difference in upstream intervention. It is true of differentiation within multi-celled animals, too. Once development with differentiation begins, differentiation can be amplified through internal feedback loops. The compartment geography of the embryo is not already present but cryptic in the egg. But the initial differentiation—telling some cells in the early embryo where they are and what to do—depends on traces generation N insert into or attach to the adapted egg (Gerhart and Kirschner 1997, Chapter 8).

If these arguments are right, even roughly, then crucial evolutionary transitions depend on the impact of organisms on the spread of phenotype variation, an impact routed through their contribution to developmental environments. To the extent that phenotype similarity across generations depends on signals sent across generations, those signals depend on channels and contexts engineered by the parental generation. If there is something to this perspective on novelty, what might it tell us about phylogenetic macroevolution? West-Eberhard has suggested that there is a critical connection between plasticity and phylogenetic macroevolution; in particular, between plasticity and adaptive radiation. For she points out that in the very same regions in which some lineages radiate, others do not. To go with the thirteen species of Galapagos finches, there is just one warbler and two flycatchers. Of these, only the Galapagos flycatcher is endemic. West-Eberhard suggests that such differences are explained by differences in phenotypic plasticity in the stem lineage. A difference in evolvability derives from differences of adaptive plasticity. Plasticity allows a lineage to simultaneously retain the capacity to change while enabling populations to specialise in the particular resources a given habitat makes available. Such populations use their pre-existing capacity to generate novelty in partnership with environmental inputs. Radiation is the response of a plastic stem species to a heterogeneous environment.

This hypothesis really does seem to fit the African lake cichlids. There is evidence that their tooth and muscle development is diet-dependent (West-Eberhard 2003, pp. 575–577). Moreover, different morphs tend to specialise in different foods when food is scare: competition seems to increase specialisation. So the spectacular radiation of the cichlids may depend on developmental plasticity in learned feeding specialisation and tooth, jaw and muscle development acting in conjunction with modularity and redundancy in their double jaw system. Likewise, the radiation of Darwin's finches may well be mediated by learning. There is evidence that these birds do learn feeding specialisations, and that those specialisations lead to the formation of subpopulations with different feeding zones, different feeding methods, and different foraging targets (West-Eberhard 2003, pp. 345–349; pp. 582–583). However, this conception of the connection between plasticity and adaptive radiation is incomplete. How do subpopulations become species?

West-Eberhard downplays the role of isolation and speciation in the evolution of novelty. She points out that significant novelties arise as variants within populations: for example, the transitions amongst insects from phytophagous life styles to carnivorous ones can begin as a within-population variation. Likewise, crucial transitions in the evolution of eusociality exist as facultative variants within populations (West-Eberhard 2003, p. 606). However, the crucial issue is not with the origin of novelties but with the genetic accommodation that makes them a more permanent, less contingent aspect of a phenotype. The second pulse on the double-pulse model of the evolution of novelty depends on isolation. For accommodation will involve the evolution of a distinctive set of gene combinations, and these would be broken up by gene flow between novel and original phenotypes. Thus the plastic stem must act in conjunction with isolation mechanisms in order to generate an adaptive radiation. One possibility is that the novel phenotypes generated by environmental induction themselves establish extrinsic barriers to gene flow. For example, some pairs of the cactus finch breed in the dry season, because they have both access to and can digest pollen. Most birds feed in the wet, relying on the insect flush for the protein they need. So exploiting a different resource can shift reproductive schedules, which would in turn limit gene flow (West-Eberhard 2003, p. 584).

It is not clear how general this mechanism is. But while a piece of the puzzle may be missing, some crucial pieces seem to have come into focus. It has been clear for some time that if the mechanisms that generate variation are biased, then those mechanisms play a central role in explaining evolutionary trajectories. Phenotypic plasticity connects phenotypic variation to environmental change, and thus shows that the mechanisms that generate variation are indeed biased. Amongst those mechanisms are the activities of organisms themselves, as they explore their world, choose and engineer their environments, and structure the context in which inherited resources are activated.

Chapter 8
The Evolution of Complexity

Mark A. Bedau

The evolution of the biosphere exhibits a trend of increasing complexity of the most complex organisms. Even though we are uncertain about the proper way to measure complexity, it is hard to deny that the earliest prokaryotic cells were simpler than the eukaryotic cells that arose from them, and these were simpler than the multicellular life forms that evolved from them, and so on. But it is controversial how to interpret and explain this trend, and even how to describe it properly (see e.g., Gould 1989, McShea 1991, Gould 1996, Sterelny and Griffiths 1999). Some think that increasing maximal organism complexity has for all intents and purposes been explained by Darwin, Miller and Urey, Watson and Crick, or some other past or present biologists. This chapter defends the alternative view that the trend is *not* yet adequately explained but instead is one of the major remaining challenges in understanding biological phenomena.[1]

My argument will crucially employ a new explanatory scheme in the life sciences: what I will call "constructive" or "emergent" computer models, which are characteristic of "soft" artificial life. These models are useful in part because they can provide clear and decisive refutation of various hypotheses about how to explain the trend of increasing organism complexity. Even more important, in general only these models can provide clear and compelling explanations for the characteristic behavior of complex adaptive systems like the evolving biosphere. Thus, these models turn out to be exactly the right kind of tool for exploring how to explain the trend of increasing complexity; furthermore, they do so in a way that is readily subject to extension and empirical corroboration. However, these models are no panacea. They can be misleading or misinterpreted, and they have limitations. As we will see when we examine specific examples, the proper use of constructive models and the proper interpretation of their behavior requires care and experience, and creating especially insightful models is extremely challenging.

M.A. Bedau (✉)
Professor of Philosophy and Humanities, Reed College, Portland, OR, USA
e-mail: mab@reed.edu

[1] Stuart Kauffman (2000) has done as much as anyone to call attention to this question. See also Farmer and Belin (1992) and Holland (1995).

A. Barberousse et al. (eds.), *Mapping the Future of Biology*, Boston Studies in the Philosophy of Science 266, DOI 10.1007/978-1-4020-9636-5_8,

8.1 The Arrow of Complexity Hypothesis

It is useful to distinguish three things: (i) a *trend*, which is simply an observed directional change in some variable in some evolving system; (ii) a *robust regularity*, which is a generic or non-accidental trend (perhaps with exceptions), a statistical "law" about the time dynamics of some variable; and (iii) a *mechanism or process* that explains a trend, whether accidental or robust. So far we have simply been observing that there is a trend of increasing complexity of the most complex organisms. This observation does not necessarily imply that the trend is a robust regularity; it might simply be an accident. Nor is any particular mechanism or process implicated as the explanation of that trend. Our starting point is merely the observation of a trend.

But perhaps the trend is not just an accident. Perhaps it is an instance of some general regularity. The *arrow of complexity hypothesis* is the hypothesis that evolution inherently creates increasingly complex adaptive organisms. This hypothesis is about the increasing complexity of the *most complex* organisms, not of all organisms or of life's mean complexity. (A closely related hypothesis pertains to higher levels of organization like ecosystems, but that is another topic.) Furthermore, the hypothesis concerns not mere complexity but *adaptive* complexity, that is, complexity that is an adaptation and serves some function. I take an adaptation to be a trait that comes to exist and persists in a lineage because it is beneficial for the organisms in the lineage; this can be cashed out by saying that the trait is produced by natural selection for that trait (Bedau 1991, 1992). This view is in the spirit of the familiar etiological or historical approach to understanding biological functions (see e.g., Allen et al. 1997).

The trend of increasing maximal organism complexity illustrates a clear and dramatic form of the emergence of biological complexity over time. Emergence had a checkered history in twentieth century philosophy because it was often construed as conflicting with forms of reductionism that were taken to be characteristic of naturalism. But notions of emergence that are compatible with naturalism and at least certain uncontroversial forms of reductionism have now been developed (e.g., Wimsatt 1997, Bedau 1997, 2003b, Boogerd et al. 2005). One would presumably want some such notion to characterize the emergence of increasing organism complexity because that trend is presumably a fully naturalistic process. The arrow of complexity hypothesis states that this dynamical emergent process is a generic property of some class of evolving systems, rather than just an accident.

The notion of complexity is notoriously difficult to define.[2] I will not propose any substantive definition here. When I talk about increasing complexity, I mean *qualitative* complexity; an organism with 100 vertebrae seems quantitatively more complex than one with 10 vertebrae, but it does not seem qualitatively more complex. An organism's qualitative complexity seems to be something roughly like the number of different kinds of fundamental properties it possesses. A vast number of different

[2] For a nice review of some of the conceptual problems, see Sterelny and Griffiths (1999).

definitions of complexity have been proposed (e.g., Edmonds 1996, Lloyd 2001, Wolfram 2002). Some of the variety reflects different attempts to capture essentially similar intuitions about complexity, but many definitions attempt to capture fundamentally different intuitions about what complexity is. I will not attempt to navigate this conceptual thicket here, but will instead for simplicity rely on a vague but not wholly unintuitive notion of qualitative complexity of organisms. Since there are different notions of complexity, there are also correspondingly different hypotheses about the arrow of complexity. Similarly, there are a variety of analogous arrow hypotheses that concern not complexity but trends in increasing disparity, diversity, body mass, and energy flow, among other notions. I will not here address whether the argument in this chapter generalizes to these other arrow hypotheses.

The arrow hypothesis should be understood as a robust regularity that has exceptions. We will see examples of evolving systems for which the arrow hypothesis fails altogether. Further, those systems for which the hypothesis holds are usually not always increasing in maximal organismal complexity; rather, sometimes their maximal complexity temporarily falls, for one reason or another. The hypothesis is that maximal complexity generically tends to continually rise. You could put it this way: For any time t with measured complexity c, there is some later time $t^* > t$ with greater complexity $c^* > c$. Any actual evolving system that we measure (such as the biosphere) at some finite time will by necessity be consistent with both the truth and falsity of the arrow of complexity hypothesis.

In addition, the hypothesis holds for evolving systems only under certain constraints, such as a continual supply of energy or material resources. There is no implication that maximal complexity must increase if those constraints fail. But the implication of the hypothesis is that the biosphere's increasing maximal complexity is not just an accident. The hypothesis does not specify the mechanism behind this (purported) robust regularity. It is rather analogous to the second law of thermodynamics, which simply states that there is an arrow of increasing entropy but does not specify the mechanism behind it. The key question about the arrow of complexity hypothesis is how to test whether it is true and, if it is, how to discover the mechanism behind it.

8.2 Replaying the Tape of Life

This issue of the creativity of evolution has been addressed by a number of biologists and philosophers. In *Wonderful Life* (1989), S. J. Gould proposed a famous procedure for testing things like the arrow of complexity hypothesis. He called the test "replaying the tape of life." To conceive of this test, one imagines a tape of the history of life, and one imagines rewinding and erasing that tape to some point in the early evolution of life. Then one imagines replaying the tape forward again, but this time having the course of life affected by different historical contingencies. In this way, what is accidental in the history of life might turn out differently, while what (if anything) is essential will remain the same. If one repeats this rewind-and-replay operation again and again and compares the results, robust regularities will stand out as statistical laws. Replaying the tape of life would be an excellent way to settle whether the arrow of complexity hypothesis is true.

Gould suggests that replaying life's tape would disprove the arrow of complexity hypothesis. He emphasizes that the evolution of life is a contingent historical process and is thus not law-like. He says that "almost every interesting event in life's history falls into the realm of contingency" (1989, p. 290), so that "any replay of the tape would lead evolution down a radically different pathway from the road actually taken" (1989, p. 51). The consequence is that all forms of life might have forever remained simple. But there is a flaw in this reasoning. Contingent historical processes can be governed by laws, as the second law of thermodynamics proves. So the fact (and it presumably is a fact) that life is a contingent historical process is entirely consistent with the arrow hypothesis. The sign of this would be that replaying the tape many times and pooling the results would reveal a robust pattern of increasing maximal complexity. Even though the details from different runs would vary due to different historical accidents, the overall pattern of increasing maximal complexity would still obtain. How could Gould have missed this? Perhaps he did not actually attempt to replay the tape of life but instead was content just to imagine what would happen.

In *Darwin's Dangerous Idea* (1995), D. C. Dennett is confident that the arrow of complexity hypothesis is true. On analogy with so-called "forced moves" that are obvious to any decent chess player, Dennett argues that complexities such as tools and language offer such obvious adaptive advantage that the stochastic process of natural selection would almost inevitably discover them. The forced move mechanism is a relevant consideration, and under certain conditions it would predict a robust regularity of increasing maximal complexity of life. But it is unclear whether forced evolutionary moves really do vindicate the arrow hypothesis. For one thing, it is unclear whether the relevant forms of complexity really do offer an appropriately large adaptive benefit over the alternatives. After all, simplicity provides its own benefits, and the final weighing of alternatives is highly sensitive to environmental context. How much confidence in the advantages of complexity is really warranted? Furthermore, even if complexity were advantageous, it is unclear whether natural selection really could discover them in a feasible amount of time, or even at all. Even if we can imagine adaptations that we could prove on engineering grounds would be clear adaptive winners, there is no guarantee that evolution can discover them by simple genetic manipulations of existing forms of life. Not even natural selection can turn every sow's ear into a silk purse. How could Dennett have been so confident in the arrow of complexity? Perhaps like Gould he did not actually attempt to replay the tape but only imagined what would happen.

Other biologists and philosophers have said things that would question or explain the arrow of complexity hypothesis, but in general they are just verbal speculations like those of Gould and Dennett. It is fine to imagine what would happen if the tape of life were replayed, provided one does not place excessive weight on the results. Complex systems are inherently unpredictable (e.g., Wolfram 1994), so imagination is notoriously faulty about their robust properties. Furthermore, the evolving biosphere is about as complex a system as there is. A central lesson of the study of complex systems is that their emergent properties can usually be discovered only by actually observing them.

But how could we actually replay the tape of life? After all, we have evidence about just one evolution of the biosphere (barring extraterrestrial travel and wonderful luck) and we cannot actually replay that tape. Now and then nature provides surrogates for replaying the tape, as when the biosphere recovers after a mass extinction, or when an isolated island is newly created and life first populates it, or when we discover a complex ecosystem that has been largely isolated from the rest of the biosphere. But there are obvious problems with these natural experiments. Too many pre-existing life forms survive the mass extinctions and contaminate the "new" biosphere, isolated islands are repopulated by existing, highly evolved forms of life, and the isolated ecosystems are not isolated enough. The obvious solution is to replay a mathematical model of an evolving system. If the model is simple enough, we can calculate its generic properties *a priori*. If the model is at all complex, we simply resort to the next best thing and simulate it on a computer. Observing the effects of smoothly sweeping through parameter space then reveals the model's generic behavior. This procedure produces hard evidence rather than wishful thinking. "Putting your model where your mouth is" in this way replaces verbal speculation with computer simulations and enforces reference only to explicit and feasible mechanisms (Bedau 1998).

8.3 Complexity Growth by Passive Diffusion

Dan McShea's models of random phylogenies (McShea 1994, 1996; see also Gould 1996) contain a population of species, and each species has a number representing the complexity of the organisms in the species. The complexity of a daughter species's is like its parental complexity, but mutations change daughter species complexity. Over time, extinction and speciation events change the composition of the species in the population, and the population evolves. A key feature of these models is that speciation and extinction events are completely random and unbiased with respect to complexity. The species that go extinct or speciate are selected from the population irrespective of their complexity. A species' complexity does not univocally affect its probability of speciating or going extinct. As a result, the model biosphere passively diffuses through the space of possible species. The models ordinarily assume that the earliest species were among the simplest possible and that there is a "left wall" of minimal complexity for any possible organism. Given these constraints (and various assumptions about rates of speciation and extinction), one can generate random phylogenies and thus replay the tape to one's heart's content. Figure 8.1 shows six examples of replaying such a model.

The first thing to appreciate is that this model seems to vindicate the arrow of complexity hypothesis. That is, a robust regularity of this model is that the complexity of the most complex organisms increases over time. There can be exceptions,

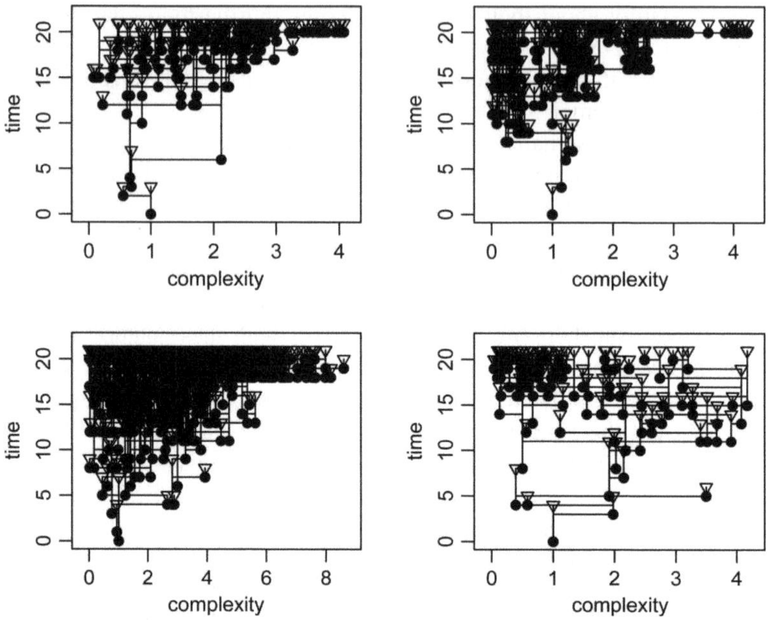

Fig. 8.1 Four examples of the phylogenies produced by a random diffusion model of the type discussed by McShea (1994, 1996) and Gould (1996). Time increases from the *bottom* to the *top*, and complexity of the organisms in a given species increases to the *right*. There is a *left* wall of complexity (complexity is constrained to be positive). Note that in general the complexity of the most complex forms of life inevitably increases over time, due to passive diffusion. That is, if you wait long enough, you will see new increase in maximal organism complexity, from passive diffusion

but they tend to be transient. The mechanism behind this robust regularity is simply diffusion in the space of recipes. No matter what the complexity distribution of the initial species is, random speciation will robustly generate more and more complex species over time.

McShea himself emphasizes a different point: Since a trend of increasing maximal organism complexity could be due simply to passive diffusion, a trend of increasing complexity does not show that complexity provides any adaptive advantage in general. Gould echoes this point in *Full House* (1996), a successor to *Wonderful Life* (1989). My concern here is with a slightly different question: What if anything do passive diffusion models show about the arrow of complexity hypothesis?

I would argue that this kind of model fails to explain the trend of increasing maximal complexity. To appreciate this, notice that a species's "complexity" in this model is entirely nominal. That is, a species does not have any real features that actually have any real degree of complexity. Setting the intended interpretation aside and simply looking inside the model itself, each species has the same complexity—they are each simple data structures with one numerical property. That property can be interpreted as complexity, but it could equally well be interpreted as any other scalar value. For example, the property could be interpreted as the species's degree of intelligence, or as the mean mass of its adult members, or as its entire biomass, or as the mean number of cell types of its members, or as its geographical range, or as the number of limbs of its members. Thus, the model does not really explain differences in complexity. It *presumes* an indefinitely large space of possible species that is structured by a relative complexity relation and that affords unlimited accessible complexity increases. It is true that diffusion will tend to increase maximal complexity given such a space, but this begs the question of whether this space exists, what constructs and structures it, and how the process of evolution enlarges those sub-regions that are accessible to further evolution. Without answers to those questions, any explanation of how complexity increases is excessively shallow.

Here is another way to see the problem. If these models really explained how complexity increases, then reinterpreting "complexity" as some other property P would produce an equally good explanation of the robust regularity of increasing maximal P in the evolution of life. To take a few extreme examples, we could interpret P as intelligence and have an argument for inevitable increasing maximal intelligence. Or we could interpret P as moral sensitivity and explain inevitable moral improvement as a robust regularity. Or we could relabel the X-axis as the number of major evolutionary transitions that have occurred and explain the major transitions in evolution as a result of passive diffusion. Obviously, any such explanations would be excessively shallow at best. Even if such arrows existed (and it is not clear that they do), these "explanations" beg all the interesting questions and make the evolutionary robustness of a trend depend on the prior presupposition that the right space of accessible evolutionary possibilities exists. When and only when such spaces exist will passive diffusion produce robust increasing trends. So the critical issue is whether and why such spaces exist and can be captured in a model. Passive diffusion models simply assume the affirmative.

Passive diffusion models do make one important point: If there is a predefined genotypic space of organisms of varying "complexity," then passive diffusion can passively explore all accessible regions of all dimensions of this space. So, the problem of modeling a trend like the arrow of complexity reduces to the problem of creating the appropriate predefined genotype space. That problem remains unsolved.

8.4 The Evolution of Complexity in Artificial Life Models

This problem would be avoided with a richer model containing entities with properties with the right intrinsic meaning. What would such a model look like? If one is interested in the arrow of complexity hypothesis, one might ask for an evolving population of entities with intrinsic properties that locate them in an infinite space of genetic possibilities, and that give them some natural intrinsic complexity. One would like the population's exploration of genetic space to be governed by natural selection operating with an endogenous fitness function that dynamically and unpredictably changed as the entities co-evolve. One would like the organisms to co-construct their effective environment, rather than merely reacting to a pre-existing and fixed environment. Since this is a computer model, one could even ask that the system itself be computationally universal, in the sense that any precisely describable process could be modeled in the system, in order to not limit the system's evolutionary potential. Many people have thought that the combination of these features should be sufficient to generate a trend of increasing maximal complexity. The good news is that a number of such models are studied in "soft" artificial life ("soft" because it is based in software, by contrast with "hard" artificial life which concerns robots and other hardware, and "wet" artificial life which occurs in a wet lab; see Bedau 2003a). These are *constructive* models in the sense that they involve creating a system that actually possesses, rather than merely can be interpreted as possessing, the properties of interest. They are also *emergent* models in the sense that, even though we have complete knowledge of the models' underlying rules and their initial and boundary conditions, it is impossible to predict or derive their global behavior—such as any trends concerning organism complexity—without observing their behavior unfold over time (Bedau 1997).

The most famous artificial life model with all of the above properties is Tierra (Ray 1992). This consists of a population of self-replicating computer programs that compete for space and CPU time in computer memory. They are loosely analogous to an evolving population of RNA strings in a chemostat. The program is typically seeded with an ancestral self-replicating program that is designed by hand. (Tom Ray's original ancestor was 90 lines of computer code.) Programs are each allocated CPU cycles, and they eventually make copies of themselves in memory. When memory comes close to getting full, older programs are removed from the system to make room for new programs. Random errors ("mutations") are introduced into programs from time to time, so that the population of programs varies. Most mutations break a program so that it is no longer able to make copies of itself, but once in a while a mutation is beneficial and allows a program to reproduce more quickly. This program will tend to produce children faster than others and so will spread through the population. Tierra became famous because a complex ecology of different kinds of programs evolves spontaneously, without having been anticipated or designed in advance by the programmer. This ecology includes parasites that take advantage of the copy instructions in their neighbors, hyper-parasites that evolve a defense against such parasites, mutualists that reproduce only in a spatially-localized colony of similar programs, and cheaters that exploit such colonies.

For present purposes, it is critical to appreciate that Tierran "creatures" (i.e., self-replicating programs) have an intrinsic rather than merely nominal complexity. For example, the 90-instruction ancestral program has an explicit copy loop that the 40-instruction parasites lack. Even without a precise definition of complexity, it is obvious that the ancestor creature is more complex than the parasite creature. It is also critical for present purposes to appreciate that Tierra does not beg but squarely faces the right questions. It has an actual space of evolutionary possibilities—the infinite universe of programs that can be written in the Tierran computer code. The actual population of Tierran programs residing in memory at any given time inhabits a small subset of that space. This space of possibilities has an intrinsic structure with respect to complexity; different programs are intrinsically more or less complex than others. Moreover, the range of possible mutations determines which nearby regions of genetic space are accessible to others. Evolution changes the accessible subspaces by changing what subspace is currently inhabited. For example, once the population has discovered a certain kind of adaptation such as unrolling the loop (Bedau and Brown 1999), further variants on that adaptation are accessible. More interestingly, since some kinds of creatures depend on interactions with others, some regions of genetic space cannot be inhabited unless certain other regions are simultaneously inhabited. For example, parasites cannot survive without hosts to exploit, and cheaters cannot survive without mutualists to exploit. Thus, the Tierra model allows us to study when, why, and how evolution creates complex "creatures."

So, how creative is evolution in Tierra? In particular, does Tierra vindicate the arrow of complexity hypothesis? The short answer is "No." Despite the complex ecology that spontaneously arises, as far as anyone knows Tierra fairly quickly stops producing anything new. The players in the ecology described above might change, but the population always consists of essentially the same kind of creatures. The production of qualitative novelty peters out below a complexity ceiling, contrary to the arrow of complexity hypothesis. This behavior of Tierra does not disprove the arrow hypothesis, of course. Rather, it suggests that Tierra is an inappropriate model for testing the hypothesis. Tierra fails to produce a trend of increasing complexity even though it encompasses creatures that co-construct their environment and evolve in an infinite genetic space by natural selection with an endogenous fitness function fluctuating via co-evolutionary dynamics, and even though the Tierran language is computationally universal. So, Tierra is a *counterexample* to the sufficiency of those mechanisms for producing an increasing maximal complexity trend. Maynard Smith and Szathmáry (1995) have said that the theory of natural selection does not predict that organisms will get more complex. Study of Tierra shows that the same can be said when we add creatures that construct their environment while evolving in an infinite genetic space according to endogeneous fitness criteria that are unpredictably buffeted by the process of co-evolution. This is negative news, but it is constructive progress because it motivates the search for new and better models.

Tierra is not an isolated example. One problem or another besets all other attempts to use contemporary models test the arrow of complexity hypothesis. For example, consider John Holland's Echo model (Holland 1992, 1995). An Echo world consists of a collection of sites, with agents migrating from site to site. The agents need resources to survive, and their activity is all geared around continually finding more of these necessary resources. Different letters of the alphabet represent different types of resources available in the world. A fixed amount of resources is distributed to each site at each time step. Each Echo agent's genetic material (its "chromosome") is composed out of the resources. The chromosome has eleven substrings; these constitute an agent's "tags" and "conditions" together with a genetic specification of which resources the agent can ingest from the environment. An agent's tags are "external" in the sense that other agents have access to them, while an agent's conditions are "internal" in that they are inaccessible to other agents. The tags and conditions are used to determine the outcome of the three types of interactions that Echo agents can engage in—combat, trade, and mating. Whether two agents interact and what type of interaction they have is determined by comparing the agents' tags and conditions. External tags and internal conditions allow complex non-transitive relationships to exist. Agents gain resources from their environment and lose them through a metabolic tax. They also can gain and lose resources by interacting with other agents.

Holland's Echo model contains a number of plausible ideas, perhaps most centrally the suggestion that evolution fundamentally affects how agents in a population interact, and it was designed explicitly to illustrate the type of complex adaptive system that could exhibit spontaneous complexity growth (Holland 1995). So it would be natural for someone to suspect that Echo would show the growth of complexity. However, as it happens, the version of Echo that is implemented today (Hraber et al. 1997) fails to show any signs of this kind of behavior (Smith and Bedau 2000). It is worth pointing out that this version of Echo is a simplified model that fails to implement some of Holland's original suggestions, so it might turn out that Holland's original hypotheses about complexity growth are still true. However, since the behavior of these complex adaptive systems is emergent, there is no way to tell whether Holland's suggestions are true until someone implements them and then empirically studies the resulting model.

Another model that has produced some impressive behavior is Avida. This model is based on Tierra but includes some interesting modifications, most notably the following three: (i) The space in which creatures interact is two-dimensional, which is more realistic than the one-dimensional spatial structure in Tierra. (ii) Creatures in Avida are typically prevented from reading the code of their neighbors, so the informational parasitism that drove much of Tierra's ecological dynamics is blocked. (iii) Creatures in Avida have the ability to input and output numbers and can garner extra energy from their environment by using these information channels to perform certain logic functions. Replaying the tape with Avida has produced some quite interesting results. Notable among these is an explanation of the evolutionary origin of complex organismal features (Lenski et al. 2003). This study demonstrated in concrete detail how natural selection can create complex features by modifying existing structures and functions.

Although the authors of this study were not intending to address the arrow of complexity, their results nevertheless illuminate our question. Avida does not vindicate the arrow, for a variety of reasons. For one thing, the complex features (certain logic functions) that evolved in Avida were explicitly anticipated by the model's designer, and the steps toward their creation were explicitly rewarded. Moreover, the "complex" features are certain logic functions that can be created out of simpler logic functions, and earlier populations were explicitly seeded with those simpler logic functions. So this result does not qualify as the kind of qualitative complexity growth that we are seeking to understand. Finally, as with Tierra, the complexity growth in Avida always reaches a plateau and stops. Thus, these results with Avida do not capture the kind of continual growth in qualitative complexity that we can observe in the biosphere, which is what the arrow of complexity hypothesis seeks to explain.

There is another study using Avida that apparently does directly address the arrow of complexity hypothesis (Adami et al. 2000). The authors investigate whether it is possible to make a case for a trend in the evolution of complexity in biological evolution, and in part because it is tractable, they focus on genomic complexity, which is reflected in the number of essential loci in a genome (i.e., the loci that would be lethal to alter). After replaying the tape numerous times, the authors conclude that genomic complexity is forced to increase, as the genome stores increasing amounts of information about the organism's environment. Interesting as these results are, though, they do not actually shed light on the arrow of complexity. As with the earlier Avida result, there is no increase in qualitative complexity. The genetic possibilities are predefined and static, so natural selection is simply fixing existing loci. Furthermore, the system is simply adapting to a pre-existing environmental complexity. The environment is fixed throughout the course of these experiments, so they do not explain how evolution creates new environmental complexity. And finally, as before, whatever complexity growth there is, it soon stops. So once again, Avida does not capture the kind of continual growth in qualitative complexity that we seek to understand.

There are a number of other interesting developments that bear on our question about the arrow of complexity hypothesis. Some developments are verbal or analytical mathematical theories, and others are constructive models of biological evolution. My remarks about them would be like those above. My general response to the merely verbal proposals is that they will remain inconclusive until they follow the maxim to "put your model where your mouth is." On the other hand, the mathematical and constructive computer models of evolution that I have seen to date all tend to suffer fates similar to those we encountered with passive diffusion models, Tierra, Echo, or Avida (Bedau et al. 2000). The key point is that none has yet produced a convincing example of the growth of organism complexity.

My diagnosis is that present models do not allow evolution continually to create opportunities for qualitatively new kinds of adaptations. Sterelny and Griffiths (1999), following Maynard Smith and Szathmáry (1995), suggest that the mechanisms of evolution have changed in ways that open up new evolutionary possibilities. What is missing in present models, I think, is the opportunity for the mechanisms of evolution to change in this way—and especially for it to happen not by some *ad hoc* device but as a result of the process of evolution itself. This is not impossible in principle, but it is missing in practice in today's models. The key question, then, is to determine what mechanisms would suffice to achieve this kind of qualitatively expanding evolutionary process.[3] I should emphasize that my diagnosis is just a hunch. People used to have other hunches, e.g., that an infinite genetic space explored by a dynamic endogenous fitness function would achieve significant

[3] No doubt niche construction (Odling-Smee et al. 2003) will figure significantly among the mechanisms in a model of a qualitatively expanding evolutionary process, but it is important to point out that this mechanism is far from sufficient. For Tierra and most other artificial life models have organisms that can continually reconstruct their own environment as they evolve, and thus can continually modify the very environment to which they are adapting. Yet these models are not qualitatively expanding evolutionary processes.

evolutionary creativity (Packard 1989), but Tierra showed that such hunches were false. My hunch might suffer the same fate in the end.

The bottom line contains both good and bad news. The bad news is that today we have no model that produces a trend of ever increasing maximal organism complexity. No existing model is sufficiently creative to even test the arrow of complexity hypothesis. The good news is that soft artificial life provides a straightforward research program of constructively testing new proposed mechanisms (like the one in my diagnosis above) with new computer models.[4] I do not want to give the misimpression that this research program is easy. It can be difficult to construct a model that captures one's hunches about the key mechanisms behind evolutionary creativity. A perfect example of this is my own hunch described above. The proper conclusion to draw from this is that I do not fully understand the mechanism proposed in my own diagnosis above. For if I did, I would immediately make a model in which evolution continually creates opportunities for qualitatively new kinds of adaptations. My failure to do so to date illustrates my earlier point about the importance of moving beyond mere words and putting your model where your mouth is. It is only if and when I succeed in putting my model where my own mouth is that we will learn whether my diagnosis above has real merit—or even what exactly my diagnosis really means.

8.5 Objections and Replies

My thesis about the trend of increasing complexity and the arrow of complexity stands in need of some clarifications and qualifications. A good way to develop them is by reply to objections that might be raised against my argument.

Self-sealing argument. My argument might seem self-sealing for the following reasons. I argued that Tierra was an inadequate model for our purposes on the grounds that it did not produce a trend of increasing organism complexity. Any model adequate to test the arrow hypothesis must be able to produce this trend. But then won't the model necessarily confirm the arrow hypothesis? If a model fails to confirm the hypothesis, won't it be ruled inadequate like Tierra? If so, then it is impossible to disprove the arrow hypothesis and my argument is self-sealing.

This worry is based on a misunderstanding of replaying the tape of life. The issue is whether the observed increase in complexity is merely an accident, or whether it is a robust feature of systems like the biosphere. Since we cannot actually replay the course of evolution, we seek to do the next best thing and replay a model of reality. When looking for adequate models to test the arrow hypothesis, we must consider three questions. The first is whether the model can produce a trend of increasing

[4] Peter Godfrey-Smith claimed that progress in artificial life has "ground to a halt" (Godfrey-Smith 2003). My argument about using artificial life models to explain the evolution of complexity suggests otherwise. Within the artificial life community the evolution of complexity is a grand challenge, one that might take many years of sustained effort to solve. In general, artificial life has been making stready progress on a diverse portfolio of projects (see the recent review in Bedau 2003a)

maximal organism complexity, the second is whether the model is like reality in the relevant respects, and the third is whether the model produces the trend robustly or merely as an accident. If a model like Tierra cannot exhibit a trend of increasing complexity, then the model should not be used to replay the tape. The biosphere *does* produce the trend, so the model's inability to reproduce that behavior disqualifies it as an adequate model of the biosphere. On the other hand, if the model can produce the trend, then it remains a candidate for replaying the tape.

Our next question is whether the model is like the biosphere in the relevant respects. Since the point of experimenting with models is to discover what would happen if we could replay the course of evolution, the models must capture the key features of reality. For example, they should presumably exhibit the process of natural selection operating on an indefinitely large space of genetic possibilities. Of course, the models will simplify away a great number of properties of the actual biosphere, so they will purposely be unrealistic in those respects. Exactly which features of reality are crucial to capture in an adequate model is one of the open questions that we hope to answer by our attempts to replay the tape, and the proper answer to this question will be disputed (see below). But there are still plenty of features that will clearly disqualify a model. For example, we should disqualify any model that explicitly gives an *ad hoc* fitness boost to organisms that is proportional to their complexity, for this model would beg the very question under investigation. (Some versions of Avida use a fitness function that explicitly rewards organism complexity, as we noted above.) Similarly, we should presumably dismiss any model that assumes that all complexity increases are simply random mutations on a par with all other random mutations (as simple random diffusion models do).

If a model can produce the trend of increasing complexity and it seems realistic in the relevant respects, then it is a plausible candidate for replaying the tape. The key question now is whether the trend of increasing complexity is a robust feature of the model. Note that this is an open question. There is no guarantee that a model that can produce the trend will do so typically or generically. Those like Gould in *Wonderful Life* who think that the observed trend of increasing complexity is mere contingency should expect the same kind of behavior from the model. It is entirely possible that when the model produces a trend of increasing complexity, this is only an accident. Thus, my argument is not self-sealing. What matters is whether realistic models that can produce a trend of increasing complexity do so typically.

For the sake of simplicity, I am purposely downplaying certain complications that might arise when replaying the tape. Eventually we might have two realistic models that can produce the trend, but the trend might be robust in only one. In this and similar circumstances, we would need to take a closer look at what explains the difference between the models and how the models compare with what we know about reality. We might then choose to frame more detailed hypotheses and test them. All of this could be done in a constructive and empirically grounded manner. But first things first. Before we worry about this hypothetical situation, we must deal with our actual situation and find any relevantly realistic model that produces a trend of increasing organism complexity.

Saved by computational universality. A number of artificial life models like Tierra are computationally universal (Maley 1994); that is, they have the power of a universal Turing machine and can be programmed to compute any algorithm or simulate any mechanical process. For any such model, it might be thought to be trivially, even mathematically, true that the model can produce a trend of increasing complexity. For surely there is a mechanical process that produces such a trend; after all, the biosphere is exhibiting the trend and it surely is a mechanical process, at bottom. So any computationally universal system would be capable of producing that trend. Thus the arrow of complexity hypothesis would be trivially true.

There are two reasons we can safely dismiss this worry. The first we canvassed above. Even if a system can produce a trend of increasing complexity, it is still an open question whether that trend is robust. So, far from being trivially true, the arrow of complexity hypothesis might still be false for the model in question. The second reason not to think that computational universality trivializes our question involves distinguishing what is possible in principle from what is possible in practice. The argument from computational universality shows only that producing the trend of increasing complexity is possible in principle. But what is relevant here is a model's ability to produce the trend in practice. The biosphere is actually producing the trend, so obviously it can produce it in practice. Thus, a model that can produce the trend merely in principle but not in practice is unlike the biosphere in the relevant respect and so is unsuitable for replying the tape. And of course, plenty of things that are possible in principle are impossible in practice. So computational universality is a red herring. It is not sufficient for showing that a model will vindicate the arrow of complexity hypothesis, nor even for showing that a model can in practice exhibit a complexity growth trend.

Conway's famous Game of Life (Gardner 1983, Poundstone 1985) nicely illustrates the irrelevance of what is possible merely in principle. This "game" is "played" on a two-dimensional rectangular grid of cells, such as a checker board. Time is discrete. A cell's state at a given time is determined by the states of its eight neighboring cells at the preceding moment, according to the following birth-death rule: A dead cell becomes alive if and only if 3 neighbors were just alive, and a living cell dies if and only if fewer than 2 or more than 3 neighbors were just alive. (Living cells with fewer than two living neighbors die of "loneliness," those with more than three living neighbors die of "overcrowding," and a dead cell becomes alive if it has the three living neighbors needed to "breed" a new living cell.) The Game of Life is a cellular automaton—the sort of system that Wolfram has promoted as a paradigm of complex systems (Wolfram 1994, 2002).

Some Game of Life initial configurations remain unchanging forever others oscillate indefinitely still others continue to change and grow indefinitely. One simple and striking example—dubbed the "glider"—is a pattern of five living cells that cycles through four phases, in the process moving one cell diagonally across the Life field every four time steps. Some other notable patterns are "glider guns"—configurations that periodically emit a new glider—and "eaters"—configurations that destroy any gliders that collide with them. Clusters of glider guns and eaters can function in concert just like AND, OR, NOT, and other logic gates, and these gates can be connected into complicated switching circuits. In fact, Conway proved (Berlekamp et al. 1982) that these gates can even be cunningly arranged so that they constitute a universal Turing machine, and hence are able to compute literally every possible algorithm, or, as Poundstone vividly puts it, to "model every precisely definable aspect of the real world" (Poundstone 1985, p. 25). So, since the trend of increasing complexity is a "precisely definable aspect of the real world," the Game of Life can model that trend—at least, in principle. However, there is no reason to think that it can model that trend in practice. The problem is that the Game of Life configuration that constitutes a universal Turing machine would be so large, and the algorithms for modeling the trend would be so slow, that the space and time requirements for the computation make the task completely impractical.[5]

Enough space and time. At this juncture a related worry arises: Maybe we have failed to observe Tierra and similar models produce a trend of increasing complexity merely because we have failed to give the models enough memory space or enough CPU time to show what they can do. Maybe we need only give them a larger population or let them evolve longer. When one compares population sizes and numbers of generations, the space and time resources available for existing models seems woefully small compared to the biosphere. A typical artificial life model has a population size between 100 and 10,000 organisms, with the largest population sizes being on the order of 1,000,000. By contrast, the number of organisms in the biosphere is many orders of magnitude larger. Similarly, life has existed on the Earth for billions of years, and over 500 million years have elapsed since the Cambrian explosion. Even with the fastest computers today, artificial life models have not produced anything approaching these numbers of successive generations. Thus one might conclude that Tierra and other models just need comparable numbers of organisms undergoing comparable numbers of generations in order to exhibit comparable trends in complexity growth.

One way to make this worry vivid is to consider the following simple thought experiment involving complexity growth. Imagine a Petri dish covered with nutrient agar that is dotted with a number of bacterial colonies. You can also imagine continually inducing mutations in the bacteria. Now, ask whether you would expect

[5] A brief document available on the web reports the invention of a (non-universal) Turing machine implemented in the Game of Life (Rendell 2001). The device employs 1714 × 1647 cells (and uses additional cells to represent the tape). It uses 11,040 Game of Life generations to accomplish one Turing machine cycle. I am not aware whether this report has been independently verified. It is reported that the Turing machine had 3 states and 3 symbols, but that the methods could be extended to produce a 16 state and 8 symbol universal Turing machine. Such a universal Turing machine would reportedly be almost half again as slow, taking 15,360 Game of Life generations per Turing machine cycle. Representing the universal Turing machine would take more than an order of magnitude more cells (about 10^7), and storing even a very simple program for the universal Turing machine would increase the size of the initial configuration by about two orders of magnitude (10^9). A program that would implement an evolutionary process exhibiting a trend of increasing organism complexity would be many orders of magnitude larger than this. These numbers indicate the sense in which it would be completely impractical to model the trend of increasing complexity in the Game of Life, even though this is possible in principle.

to observe a trend of increasing qualitative complexity. Of course, you would not see any such trend. The bacteria might evolve a bit because of the mutations, but they would remain bacteria, and indeed essentially the same kind of bacteria. But now notice that this imaginary system is much more complex than any artificial life computer model; the Petri dish contains vastly more organisms, each one of which is itself a self-assembling and self-organizing molecular machine that is vastly more complex than the most complicated cell simulation. So, if the Petri dish system is too simple to exhibit complexity growth, how could we possibly expect it of any existing artificial life model—unless its spatial and temporal resources are vastly increased?

The first point to make in reply to this worry concerns the Petri dish thought experiment. There is an obvious sense in which the Petri dish system would be more complex than any artificial life model, because the bacterial population is orders of magnitude larger than the model population and the number of molecules in each bacterium is orders of magnitude larger than the number of bits used to represent each virtual organism. But that might not be the relevant comparison, for those might not be the properties that matter. Once population size is large enough that random genetic drift does not dominate natural selection, it is not evident that population size itself directly affects the existence of trends concerning the evolution of organism complexity. Similarly, it is not at all evident that the number of molecules in an organism reflects the complexity of the kinds of behaviors or interactions in which the organism can engage—yet this latter is presumably the kind of property that bears on whether a system will exhibit a trend of increasing organism complexity. Another property that presumably would affect such trends is the complexity of a system's niche structure: how complex each niche is, and how complexly they are all connected. But an artificial life model might well be more complex than the bacterial population in the Petri dish with respect to the properties that matter. It is relatively easy to engineer models in which virtual organisms can engage in a very large number of different kinds of behavior and interactions. Similarly, while the bacteria in the Petri dish would all experience essentially the same kind of environment, it is easy to engineer a model with an environment that has very different kinds of niches connected in a very complex topology. Thus, it is not at all clear that the Petri dish system is more complex than artificial life models in the relevant respects.

The main point to make in response to the conjecture that existing models would behave qualitatively differently if given vastly more space or time is that we simply have no compelling evidence to support this conjecture. What would such evidence look like? One obvious form of evidence would come from experiments in which spatial and temporal resources are increased incrementally. If such increases often or even sometimes did cause the models to evolve qualitatively more complex organisms, this would certainly count in favor of the conjecture. However, I am unaware of any such experiments having been done thoroughly and carefully. And the anecdotal experience of myself and my colleagues all points in the opposite direction. When we increase the space or time available to Tierra and other models, they continue to exhibit the same kind of behavior and qualitatively more complex organisms never appear. Of course, this does not prove that Tierra or other existing models cannot produce a trend of increasing organism complexity given yet *more* space and time. Today this remains an open empirical issue. It is worth emphasizing that this issue can be addressed constructively and creatively by devising and experimenting with artificial life models.

Hopelessly complicated. One might conclude from my argument not just that Tierra is missing some key mechanism and that we failed to provide it with sufficient time and space, but that the model itself is much too simple. Actual biological organisms are vastly more complicated than any virtual organism in a model, and the web of interactions that connect actual organisms in the biosphere is vastly more complicated than the networks connecting virtual organisms in a model. As biologists learn more and more of these details, we cannot help being struck by how complicated the biosphere is. No doubt many of these complications figure into the process that produces the trend of increasing organism complexity. On the other hand, there is a limit to how complicated any practical computer model can be, because of constraints on the number of lines of code that human can write and understand. One might conclude from this that no practical computer model could ever be complicated enough to capture the features that enable the biosphere to exhibit the complexity trend. Too many details about the biosphere are critical, and any sufficiently realistic model would need to be hopelessly complicated.

If this worry is sound, then we will never devise a model that is adequate to replay the tape of life. If I am right that we must replay the tape in order to evaluate the arrow of complexity hypothesis, then this worry entails that the arrow of complexity hypothesis will remain shrouded in mystery. We will never learn whether the remarkable growth in complexity that we can observe in the biosphere is merely an accident or an inherent property of evolving systems like the biosphere. However, there is no reason yet to draw this bleak conclusion. We have no good reason to think there is no simple feasible mechanism that will produce the complexity trend. Plenty of simple mechanisms have not yet been explored. In addition, plenty of more complicated models could be devised. So it is much too premature to conclude that any adequate model must be hopelessly complicated. It is another open empirical question—and one that artificial life models can again constructively explore—whether practical computer models can capture whatever it is that enables the biosphere to exhibit growing organism complexity.

Try another model. I have been emphasizing Tierra because it embodies six mechanisms—a population of creatures, represented in (i) a computationally universal language, that (ii) co-construct their environment and evolve in (iii) an infinite genetic space by (iv) natural selection with (v) an endogenous fitness function fluctuating via (vi) co-evolutionary dynamics—that many have thought were sufficient to produce a trend of increasing organism complexity. Some people might respond that the scope of my argument is limited to Tierra and that Tierra's limitations do not show that those mechanisms themselves are incapable of producing a trend of increasing complexity. They might conclude that some other model with the same six mechanisms could well produce the trend, and thus that those mechanisms would be sufficient to explain the trend. If there is a problem, it is not with those mechanisms but with Tierra.

This worry fundamentally challenges my argument, and I think it is fundamentally misguided. To see why, we need to be careful about what is at issue. First, I would agree that another model with different mechanisms might well produce a trend of increasing organism complexity, but, of course, that is not the issue. Second, I would also agree that another model with those same six mechanisms *and with some other important difference from Tierra* might well be able to exhibit the trend. But that is also not the issue, for it would not show that those six mechanisms by themselves were sufficient to produce the trend. Rather, it would show that those mechanisms *plus something else* are sufficient for the trend. Since Tierra embodies those mechanisms and yet fails to produce the trend, Tierra shows that those mechanisms together are insufficient to explain the trend. This conclusion is not limited merely to Tierra; it applies to any model that embodies those same mechanisms. I presume that we will eventually devise a model that exhibits a complexity growth trend, and I suspect that it will embody many, perhaps all, of the six mechanisms from Tierra. But it must significantly differ from Tierra; it must contain some further key mechanism.

Some might think that it is trivial to identify the key mechanism missing from Tierra. To them I repeat my earlier injunction: "Put your model where your mouth is." One of the important morals to draw from current efforts to resolve the arrow of complexity hypothesis is that talk is cheap and inconclusive. As we have seen, it is all too easy to spin a superficially plausible tale about why some mechanism "surely must" yield a trend of increasing complexity. This is exactly what people in artificial life originally thought about the six mechanisms in Tierra—and that is why they created models like Tierra that embodied those mechanisms. But remember that the global behavior of artificial life models is emergent; there is no way to tell how they will behave short of watching them unfold in time. This emergent behavior is the reason why verbal speculation about the arrow of complexity is so inconclusive, and why the acid test of any hypothesis about the arrow is to implement a model that embodies the hypothesis and study the model's behavior empirically.

The arrow is false. I have been arguing that the growth of complexity in the biosphere has not yet been adequately explained, and that the arrow of complexity encapsulates the central hypothesis to be resolved. Some might be skeptical about this project, because they believe that the arrow of complexity hypothesis is simply false. They are convinced that the biosphere is not exhibiting any robust trend that involves the generic production of increasing complexity. Rather, this is simply an accident. Thus, there is nothing interesting to explain and my project is misguided.

There are two things wrong with this worry. First, I have not been arguing that the arrow of complexity hypothesis is true, but that it is an important open question. For all we know, it might be true and it might be false. At this stage, claims that that the arrow is false are dogmatic; they are just expressions of faith. (The same holds for claims that the arrow is true.) Second, even if the arrow of complexity hypothesis does turn out to be false, there is still something important left to explain: the observed trend in increasing organism complexity. Even if that trend is an accident, it still has an explanation. So, among other things, we should be able to explain how it is possible for that kind of accident to happen.

The fact of the matter is that today we have no compelling explanation for how a trend of increasing organism complexity could happen accidentally. It is not that such an explanation is impossible, but that we do not have one in hand yet. What would constitute such an explanation? The most convincing demonstration would be a model that was arguably like the biosphere in the relevant respects and that could exhibit the trend. If we want to explain how the trend could be an accident, then we should show that the trend is not a generic or typical feature of the model. Rather, the trend should be an accidental feature of the model, but still observable. In this respect, it will differ from passive diffusion models, for they make the growth of complexity a typical trait.

Whether you want to argue for or against the arrow of complexity hypothesis, the critical first step is to produce some appropriate model that actually produces the trend. This will set in motion further steps such as investigating whether the model is like the biosphere in the relevant respects, and assuring ourselves that the model produces real and not merely nominal complexity and does not beg any questions of how the space of genetic possibilities is structured. Without this kind of constructive demonstration, professions about why the arrow of complexity "must" be true or false simply ring hollow.

Odious allies. There is one more objection that I should briefly discuss. I have been arguing that nobody yet knows how the process of evolution could have created a biosphere with extremely complicated organisms from a biosphere containing only simple organisms. This conclusion should be music to the ears of creationist and intelligent design critics of Darwinism who argue that the process of evolution cannot explain how the biosphere contains complex organisms (see e.g., Wells 2006, and for a more balanced treatment, Pennock 2001). I have not argued that evolution *cannot* explain complexity, but that it has not yet explained complexity, and the fact that we do not yet know how evolution can create complex organisms of course does not imply that evolution cannot create complex organisms. Further, I have been arguing that the appropriate artificial life models will eventually provide a sound naturalistic explanation for the evolution of complex organisms. But dogged defenders of Darwinism will worry that my argument will give aid and comfort to religious critics of Darwinism, that they will cherry pick what is useful for their purposes and twist it into support for unscientific and anti-naturalistic alternatives to Darwinism.

There are two points to make about this worry. For one thing, it is a bad idea to pretend that we can explain the evolution of complexity before we actually can, for this just supports the claims of Darwinism's critics that the evidence for Darwinism has been "exaggerated," "distorted," or "faked," as claimed on the back cover of Wells (2006). The main point to make is that it is intellectually dishonest to deny that the evolution of complexity is an open question today. That alone should settle the issue. None of us can prevent the unscrupulous from abusing or misusing what we say, but it would be sad indeed if this fear were to muzzle frank and open investigation of the limits of current scientific explanations. That would simply retard progress towards more adequate explanations of the evolution of complexity.

8.6 Conclusions

A fundamental challenge in the life sciences today is to settle whether the arrow of complexity hypothesis is true and explain the observed trend of increasing maximal organism complexity. Progress on this challenge is slowed in part by lack of recognition that this issue is still open; many people do not realize that the answer to this important question is still a mystery. Some people think that natural selection, given an infinite space of genetic possibilities, will inevitably produce more and more complex adaptations. But soft artificial life models like Tierra show conclusively that those mechanisms are in general insufficient to produce a trend of increasing complexity. The proof is simple: The models embody those mechanisms but they do not exhibit the requisite behavior. Mechanisms like natural selection in an infinite space of genetic possibilities might still be necessary for explaining the trend, but Tierra shows that they are not sufficient.

The conclusion to draw from this is that biology needs new concepts, theories, and models if it is to resolve the arrow of complexity hypothesis. The status of the arrow of complexity remains an open question today. My own hunch is that we are missing some key insight, some important new concept or process or mechanism that will resolve why complexity grows over the course of evolution. Once we discover this new idea, it might seem obvious and almost trivially true, like the idea of natural selection seems to us in hindsight. But also like natural selection, before we discover the new idea, our explanations of the evolution of complexity fail to convince, and we feel like we are groping in the dark.

Fortunately, biology does have a new powerful tool for exploring answers to this question: the constructive models of soft artificial life. These models are essential for making progress on deep questions about complex adaptive systems, such as the arrow of complexity. Because these systems are complex, their global behavior is typically emergent. Thus, the only way to be sure how such a system will behave is to build an emergent model of the system and observe its generic behavior. The process of using constructive models to investigate issues like the arrow of complexity is not foolproof, however. As we have seen, some models beg the interesting questions, and others fail to produce the appropriate behavior. So, proper use of constructive models requires care and experience. Fortunately, in the right hands, they provide a public, repeatable, and empirically grounded method for making incremental progress on deep questions about biological phenomena. Appreciating how constructive models can resolve the arrow of complexity can start to suggest how similar models can play a central role in answering other fundamental biological challenges in the next decade and beyond.

Chapter 9
Self-Organization, Self-Assembly, and the Origin of Life

Evelyn Fox Keller

9.1 Introduction

Interest in the role of "self-organization" in biological evolution and development has grown sharply over the last couple of decades, but confusion and uncertainty over the meaning of the term remains widespread. Usage has varied greatly since the term was first introduced in 1790, and a very crude taxonomy might arrange its various meanings according to three different contexts, in three different time periods, roughly corresponding to the biological, the engineering, and the mathematical/physical. Immanuel Kant first introduced the term to characterize the distinctive feature of organisms; later, in the mid 20th century, it was appropriated for a series of conferences organized addressed to the design of intelligent and self-acting machines; and finally, two decades later, it was reintroduced by physicists and mathematicians working in statistical mechanics and non-equilibrium thermodynamics for describing such phenomena as turbulence and thunderstorms, with the explicit hope that such analyses would also provide us with an understanding of the emergence of biological entities out of physical and mechanical precursors – i.e., with an understanding of the origin of life.

Yet even within each of these contexts, the meaning of self-organization remains difficult (if not impossible) to pin down in actual practice. For example, confining ourselves just to current usage, we find that the term is often used to exclude processes of self-assembly and stigmergy (especially when employed in technical contexts), but at other times we see it being used quite inclusively. I myself favor a maximally inclusive use, partly because the oppositions one sees discussed (based on exclusive uses of the term) are often based on faulty premises, and partly because of the limitations of the particular theoretical formulations in which exclusive uses of the term are rooted. I argue that these limitations become particularly evident when thinking about the role of *self-organization* in the origin of life.

E.F. Keller (✉)
Professor of the History and Philosophy of Science, MIT, Cambridge, MA, USA
e-mail: efkeller@mit.edu

A. Barberousse et al. (eds.), *Mapping the Future of Biology*, Boston Studies in the Philosophy of Science 266, DOI 10.1007/978-1-4020-9636-5_9,

For example, one distinction that has become routine in the literature is that between self-organization and self-assembly. I quote from a recent, fairly representative, article in *Science*:

> Two fundamentally different mechanisms can account for the generation of complex structures from random mixtures of small molecules. One mechanism is self-assembly near thermodynamic equilibrium. A very different mechanism is self-organization in energy-dissipating systems. Although they do not reach thermodynamic equilibrium, these systems can reach steady states (Surrey et al., 2001: 1167).

The ostensibly clear distinction that is emphasized here is that between systems at or near thermodynamic equilibrium and those far from equilibrium, where the most commonly cited example of self-assembly (the former) is *in-vitro* viral assembly, and that of self-organization (the latter), the emergence of Bénard cells or Turing patterns.[1] However, the problem with this distinction is that, while Turing patterns undoubtedly require a flow of energy, and hence a departure from TD equilibrium, so too does viral assembly. While it may be true that the net result of viral assembly appears to be a static structure, and many Turing patterns are not, the fact of the matter is that the process of self-assembly is a conspicuously dynamic one, often requiring the agency of ingenious motors working against dramatic differentials in pressure (see e.g., Chemla et al., 2005). (These motors are presumably buried in the "packaging mix" employed in the laboratory practices of i*n vitro* construction of viruses.) Furthermore, some Turing patterns can themselves result in the establishment of static structures (witness, e.g., the formation of zebra stripes). Thus, the basis of the distinction between self-assembly and self-organization that is here claimed, at least for these examples, falls away and we are left with no obvious way of separating these various processes.

Indeed, self-assembly (or, alternatively, composition) is of paramount interest to me here. It is a crucial component of Herbert Simon's argument for the evolution and hierarchical organization of complex architectures (1962). Indeed, I argue for the particular appropriateness of revisiting Simon's arguments about composition and hierarchy for thinking about the emergence of living entities. These arguments seem to me especially relevant for thinking about pre-Darwinian evolution, for they provide us with a way of understanding the operation of a cumulative evolutionary process prior to the advent of Natural Selection. Not survival of the fittest, but survival of the most stable. Or, as I would prefer to call it, survival of the most robust.[2] But before saying more about Simon's arguments, a word about another opposition that is sometimes invoked, albeit on more epistemological grounds is also in order.

For some, hierarchy and self-organization constitute opposing explanatory schemas, indeed, with opposing political valences. Despite Simon's efforts to distinguish his use of the term from conventional connotations, hierarchical construction

---

[1] *In-vitro* viral assembly refers to the process by which, given a supply of viral genomes and what in the trade is called "packaging mix", viral heads spontaneously assemble, and the DNA is, also spontaneously, packed inside. The result, amazingly, is a fully functioning virus.

[2] A growing literature attests to the importance of robustness in evolution, and Kitano (2004, 2006) provides a helpful overview.

is often associated with social hierarchy, while self-organization is understood as organization from below. I would argue, however, that self-assembly by composition should itself be seen a form of self-organization, for the hierarchy that Simon envisioned is simply the consequence of the iteration of such modes of self-organization (or self-assembly) over time. I suggest that where difference really does need to be marked is three-fold: first, between the iterative processes of self-organization that occur over time, and the one-shot, order-for-free, kind of self-organization associated with the non-linear dynamical systems that mathematicians usually study; and second, between the heterogeneity of complex systems and the uniformity of simple gases, lattices, or fluids, and finally, between multi-level architectures and horizontal structures. It is here, in these distinctions, where I believe we will find the crucial difference between the kinds of complexity found in organisms and machines, or, as some have put it, between emergent and organized complexity.

More specifically, I want to suggest that evolution by composition has particular importance for thinking about the emergence of cellularity, and especially, of that quintessentially biological attribute, function. Biological cells are replete with devices for ensuring survival, stability, robustness, and the acquisition of such a device automatically endows the system in which it arises with a selective advantage – selection for robustness. Given the emergence of simple devices, arising out of the fortuitous combination of already existing molecular complexes, it is not hard to imagine their subsequent refinement, elaboration, and integration into every more complex structures, all by virtue of the enhanced stability they would provide. But a major role for fortuitous combination would work strongly against a continuous increase of either complexity or robustness and would argue instead for the existence of major discontinuities.

9.2 The Origin of Life in Four Acts

As indeed would seem to have been the case. The historical record supports the expectation that different mechanisms for ensuring robustness marked off different evolutionary epochs, and my (admittedly selective) reading of the literature suggests three epochal divides. If we start by assuming the early existence of autocatalytic systems of some form – systems with built-in mechanisms for more-making – the arrival of nucleic acid molecules might be taken to mark the first major discontinuity. Such molecules, which almost certainly appeared on the scene long before the advent of anything like a primitive cell, introduced a significant advance over earlier mechanisms for auto-catalysis, precisely because they made possible the replication (a particular kind of more-making) of molecules with arbitrary sequences. But the presence of nucleic acid molecules does not yet imply the presence of genes. That requires the arrival of a translation mechanism between nucleic acid sequences and peptide chains, and of necessity, it must come later, for it requires the combination of already existing nucleic acid molecules AND protein structures, but that innovation – in effect, the advent of genes – ushered in an entire new order of evolutionary dynamics. During the next epoch – the few hundred million years over which cellularity evolved – change seems to have depended primarily on the horizontal flow of genetic bits between porous entities (or proto-cells) that are not yet sufficiently sealed off to qualify as candidates for natural selection. Carl Woese argues that cellular evolution, precisely because it needed so much componentry, "can occur only in a context wherein a variety of other cell designs are simultaneously evolving ... [and] globally disseminated." He writes, "The componentry of primitive cells needs to be cosmopolitan in nature, for only by passing through a number of diverse cellular environments can it be significantly altered and refined." Similarly, he also concludes, "Early cellular organization was necessarily modular and malleable" (2002: 8746).

Only with the sealing off of these composite structures and the maintenance of their identity through growth and replication – i.e., after a few hundred million years of extremely rapid evolution – did individual lineages become possible, and this marks the third major discontinuity. With individual lineages (and the predominance of vertical gene transfer), the operation of the entirely new, albeit far slower,[3] kind of selection that we call Natural Selection. [As Freeman Dyson puts it, "one evil day, a cell resembling a primitive bacterium happened to find itself one jump ahead of its neighbors in efficiency. That cell separated itself from the community and refused to share. Its offspring became the first species. With its superior efficiency, it continued to prosper and to evolve separately" (2005)]. Woese calls this the Darwinian threshold. Much of the variety we see around us is a product of Natural Selection, but most of the enzymatic reactions on which these forms are based depend on proteins and protein machinery that were already in place with the arrival of single cell organisms. Furthermore, many of the molecules and machinery responsible for more recent innovations have been formed by cutting and pasting earlier structures, or by recombining them in ways that allow them to be employed in different temporal and spatial contexts. In other words, even with the advent of natural selection, evolution by composition (or, to use a term of currently more popular, tinkering) seems to have continued to play a critical role in the creation of novelty. As Marc Kirschner and John Gerhart put it, most post-Darwinian novelty "comes about by the deployment of existing cell behaviors in new combinations and to new extents, rather than in their drastic modification or the invention of completely new ones" (2005: 39).

Probably the most important such innovation was the invention of development, the process by which higher, multicellular organisms are produced, and this innovation clearly did come about after the advent of Natural Selection. Development stands in sharp contrast to evolution in a number of ways, but the contrast between Simon's model of evolution of complex structures by assembly or composition and the development of multi-cellular organisms by differentiation and morphogenesis is particularly sharp. Simon claimed that it doesn't matter, that the end result is the same. Multi-cellular organisms, he pointed out, have the same hierarchical (or

[3] I have not discussed rates of evolution here, but one of Simon's most important arguments was that the time required for the evolution of a complex system by assembly is a minute fraction of that required for evolution by sequential accumulation of novelty (see esp., pp. 188–189).

modular) structure as that of systems that had formed by composition. It is the structure, he argued, and not its method of production, "that provides the potential for rapid evolution" (p. 193). Of course, whether or not he was right is another question altogether, but I want to leave it as just that, i.e. as a question.

Transitions in the course of Darwinian evolution are of course well known, and they remain an obvious challenge, notwithstanding the immense strides that have been made over the last couple of decades in our understanding of evolution and development. And arguably, the challenge posed by the transitions that are thought to mark the pre-Darwinian scenario, the developments that might plausibly be supposed to have led to the emergence of the modern cell, is even greater. Indeed, it is difficult to even begin to imagine an account based on a gradualist framework of evolution – i.e., on the gradual accumulation of small random changes (whatever the substrate of such changes) – that would suffice. Far more promising, I suggest, are the efforts that have recently begun to implement Simon's proposals and to model evolution by composition along a number of different lines. A recent book by Richard Watson (2006), e.g., provides a useful review of computational models of what he calls "compositional evolution" (under which he includes modular assembly, sexual recombination, horizontal gene flow, and the hierarchical encapsulation of symbionts – all processes that involve the combination of semi-independent preadapted subsystems). Just as Simon predicted, these models demonstrate a radical extension of the domain of evolvability (i.e., of what can evolve in a finite time). Perhaps of special interest for the emergence of cellularity are the models of symbiotic encapsulation that allow for the combination (and subsequent integration) of systems with non-overlapping properties. In a similar vein, Hiroaki Kitano argues for "classes of biological systems [that] have evolved to enhance their robustness by extending their system boundary through a series of symbioses with foreign biological entities." Such "self-extending symbiosis" may have particular importance for early evolution because of the ways in which increased robustness is thought to foster evolvability (2006: 61).

9.3 The Gospel of Inevitability

Others have argued for the sufficiency of a narrower conception of self-organization for these early processes, and I need to say a few words about how my view (although hardly just my own) differs. In particular, the four-part periodization I describe here, in which evolutionary history is divided by three thresholds, the nucleic, the genetic and the Darwinian, bears an obvious resemblance to the tripartite periodization proposed by Bruce Weber and David Depew (1996). They too see natural selection as a phenomenon emerging out of prior (more basic) selective processes, and they distinguish these as two different dynamics, the first of which they characterize as "physical selection" (or selection of the "stable"), and the second, as "chemical selection" (or selection of the "efficient"). For them, the burden of biotic innovation falls on the process of chemical selection, in which the units of selection are cycles of energy utilization and dissipation (e.g., autocatalytic systems) competing for energy resources. Out of such selective dynamics, we can make sense of the emergence of efficient engines of autocatalysis. Call it metabolism. Innovation here is driven by "the imperatives of self-organization and dissipation" (p. 52). They may be right, and certainly the emergence of metabolism is of the utmost importance. But for me, it is only the beginning. Indeed, I suggest that both of these processes (their "physical" and "chemical" selection) can be grouped together under the dynamic of self-organization (in the physicist's sense of the term), and subsumed under a single (albeit composite) stage of pre-biotic history, prior to the advent both of nucleic acid molecules and of genes, and especially, prior to the advent of a translation mechanism linking the world of nucleic acids with that of proteins. And by contrast with Depew and Weber, the epoch I want to mark off for particular attention is the one that comes next: after the genetic threshold, but before the Darwinian threshold. In other words, where they place the burden of biological innovation on self-organization and selection for stability of states emerging out of "the imperative of self-organization and dissipation", I am placing it on selection for stability of composite structures arising out of chance encounters (or exchanges) among and between porous proto-cellular entities. If forced to choose, my imperative would thus have to be said to be selection rather than self-organization, where by selection I am thinking of the kind of selection operating in Simon's evolution by composition,. As Harold Morowitz writes, the emergence of metabolism "is embedded in the laws of chemistry, but the reactions [in the chart of intermediary metabolism] are a tiny subset of all possible organic reactions. We must search for the pruning algorithm" (p. 76). And my worry is that the dynamics of what is usually meant by self-organization does not provide such an algorithm. But if our notion of self-organization is expanded to encompass evolution by composition, then perhaps it can.

There is also another difference between us, and it is related. This is to be found in our relations to a question that has begun to assume enormous importance in the larger social, political, and cultural sphere: How high must our estimates of the likelihood of the origin of life be in order that it qualify for a physical explanation, or in order for us to claim that the origin of life is at least potentially explicable in scientific terms? For Weber and Depew, the answer is very high, amounting to a "virtual necessity". They write, "we take it as a constraint on any successful theory of the origin of life that it will reveal the emergence of living things to be a probable, indeed a virtually necessitated, outcome, rather than a 'frozen accident'" (p. 52). This is equally a crucial assumption for Kauffman (1995), for whom an opposition between improbable and "law-like" is taken for granted, and for whom to be "natural" is to be "expected", "bound to happen", while to be "expected" is to be "at home in the universe". Thus, he writes, "If we are... natural expressions of matter and energy coupled together in nonequilibrium systems, if life in its abundance were bound to arise, not as an incalculably improbable accident, but as an expected fulfillment of the natural order, then we are truly at home in the universe" (p. 20), and a few pages later, "A theory of emergence would account for the creation of the stunning order out our windows as a natural expression of some underlying laws. It would tell us if we are at home in the universe, expected in it, rather than present despite overwhelming odds" (p. 23).

Their faith is reminiscent of Ross Ashby's, who, 34 years earlier, had argued, "In the past, when a writer discussed the topic, he usually assumed that the generation of life was rare and peculiar, and he then tried to display some way that would enable this rare and peculiar event to occur. ... The truth is the opposite – every dynamic system generates its own form of intelligent life, is self-organizing in this sense" (1962: 270). Ashby's intuition now seems almost commonplace. In recent years, this refusal of what Christian de Duve (1991) calls "the gospel of contingency" has become so widespread as to prompt Eors Szathmáry (2002) to refer to the currently accepted wisdom as "the gospel of inevitability". Perhaps the religious reference is to the point, for certainly, at least for de Duve, advocacy of evolutionary determinism is offered as an explicit alternative to both divine and anthropic determinism.

The question for me is, from whence comes either the faith in the inevitability of living entities, or the assumption that underlies that faith, i.e., that inevitability (or near-inevitability) is necessary for a naturalistic account of life? Especially, I want to ask, why would one think of an unlikely event as an unnatural one? Or perhaps the assumption is merely that only very unlikely events must be accepted as unnatural? If so, then we would have to ask, what odds are we willing to tolerate in the realm of science, and at what point do the odds become so small as to be intolerable? Where do we draw the line? At one in ten? One in a thousand? One in ten million? 100 billion? One in 10^{50}? In other words, how many lottery tickets must be issued before winning the lottery becomes a miracle? Put so bluntly, it seems clear that this question takes us well beyond the realm of science.

We are abundantly familiar with the fact that the assumption of a line beyond which naturalistic explanations must fail underlies virtually all arguments for Intelligent Design. Indeed, this is the principal assumption on which Michael Behe's arguments rest. In some formulations, the assumption is that if the probability estimates is significantly smaller than one over the age of the universe (that is, the total probability is significantly less than one), then life cannot have arisen spontaneously. In other formulations, support is sought in what is often referred to as "Borel's Law" but which of course no law – only a "rule of thumb" proposed by the mathematician Emil Borel to demarcate what he saw as a realm of "events whose probability is sufficiently small [as to] never occur" (1962). But it is not only advocates of intelligent design who take it for granted that such a line exists, somewhere. To the extent that the defense of Darwinian theory rests on refuting Behe's low estimates of the probabilities, it would appear that evolutionists themselves share this assumption. And of course, we also find it appearing again in discussions of the Anthropic Principle. That it should be so widely accepted by scientists seems to me, at the very least, surprising. Not to mention perplexing. But perhaps it is less surprising to you. Perhaps you too agree that such a line needs to be drawn, somewhere. If so, then we are left only with the question of where it is to be drawn. And here, I trust that you will find it at least somewhat surprising to seeing de Duve, Kauffman, Depew and Weber and other opponents of Intelligent Design sharing with its advocates the assumption that it is to be drawn quite so close to certainty – as Weber and Depew put it, "probable to the point of virtual necessity", and as Kauffman puts it, likely enough to be "expected". What is the basis of *this* assumption?

For some defenders of evolutionary determinism, it is said to be mandated by the very method of science. Harold Morowitz, e.g., writes, "only if we assume that life began by deterministic processes on the planet are we fully able to pursue the understanding of life's origins within the constraints of normative science" (1993: 3). And in a similar vein, he equates claims (such as that of Jacques Monod) that the origin of life requires a series of highly improbable events with the view that that process "cannot be recovered from the laws of physics" (p. 13). For Morowitz, as for Iris Fry, the assumption that the emergence of life was a "deterministic" event – i.e., not the result of "a happy accident" – is "a necessary condition for a scientific study of the origin of life" (Fry, 1995: 393). Indeed, Fry argues that, despite the continuing presence of a few scientists she refers to as "the 'almost miracle' camp", this assumption underlies and unifies the entire field of research on the origin of life.

There are many problems with this reasoning, problems that run rampant through much of this literature. Here are three: First, I suggest that a certain amount of confusion between the scientific and the expected is built into linguistic habits of exceedingly long standing. The habit of labeling proper science as "natural science" threatens to bring to our expectations of science the notorious ambiguities to which the word *natural* is subject. Most immediately relevant here is slippage between the OED's primary definition of natural as "inherent in the very constitution of a person or thing"? (OED, I, 1) and the quite different (and logically unrelated) definition the OED gives next of "normal, expected" (OED I, 2).

A second problem is the tacit reliance on a 19th century model of science as nomological, and of law as necessity. Quantum mechanics dramatically undercut the belief in a deterministic world, and few philosophers of science would defend such a conception today, yet old attitudes die hard. Indeed, the very opposition between chance and necessity, contingency and determinism, derives from this older view, and when paired with a nomological conception of science, it readily lends itself to the refutation of the role of accident as unscientific and even irrational.[4]

A third problem is the tendency to bifurcate the range of possibilities for the emergence of life under suitable physical conditions into two extreme forms: on the one hand, near-inevitability (the "deterministic" position), and on the other hand, the result of a single, highly improbable event, "a happy accident", "almost a miracle" (Crick, 1981: 88), a "decisive event [that] occurred only once" (Monod, 1971: 136). Clearly, these are not the only two options, and the effect of such an artificial bifurcation seems to me unfortunate in the extreme. For one, it works to push defenders of the naturalistic view, first, into a tacit acceptance of an assumption their

[4] See e.g., Fry's claim: "That the philosophical choice is indeed one between a natural science point of view and a creationist one becomes obvious when the 'virtually zero probability' argument is carried to its logical conclusion" (1995: 399).

opponents rely upon, and second, into a position that may well prove empirically indefensible.

My own intuition is that the "gospel of inevitability" is not only methodologically unnecessary, but that it is in fact highly implausible. Just why it has become so closely tied to arguments for the importance of self-organization in the origin of life requires looking into, but it is in my view counter-productive. It is also philosophically at odds with the growing reliance of this disciplinary approach on the methods of statistical mechanics, and hence on the relevance of a world of random distributions. Indeed, it is at the very least ironic that researchers for whom the approach of statistical physics has seemed so fruitful should feel compelled to minimize the importance of random events in the historical record.[5]

The particular search for a statistical mechanics of complex systems has become a popular one throughout the world of self-organization, and Kauffman is an exemplar. To understand the emergence of biological dynamics, he investigates randomly constructed Boolean networks for statistically likely patterns of behavior of biological interest. Just as in statistical mechanics, the tradeoff here is between the randomness of the micro-dynamics (in Kauffman's case, the construction of particular networks) and near-certain (or law-like) behavior on the macro level (most interestingly for Kauffman, catalytic closure). His Boolean networks are not intended to simulate real biological networks, but the underlying assumption is that the generation of such statistically likely patterns of behavior will inform us about the organization of actual biological systems. But this just brings us back to the original question: how likely or unlikely are the forms of organization that we find in the organisms to which evolution has actually given rise? Or, to put the question differently, given the uniqueness of the evolutionary record, is there any reason to associate the historical route to biological design that evolution actually took with a statistically likely one, even supposing one could formulate the latter?

This question, of whether biological processes belong to the realm of the statistically likely or unlikely also lies at the heart of the controversy currently raging over the significance of "scale-free" laws in biology. Albert-László Barabási and his colleagues have argued for network topologies built by random association, enjoined only by the rule of preferential attachment. But John Doyle and his colleagues counter that such topologies, although likely, perform poorly. By contrast with Barabási's topologies, topologies of the networks designed by Doyle and colleagues to optimize performance under existing constraints, are, while effective, highly improbable. Indeed, Doyle and his colleagues go so far as to conclude that the "likely" topologies "have such bad performance as to make it completely unrealistic that they could reasonably represent a highly engineered system..." (Li et al., 2004).[6]

---

[5] See Fry p. 413, in relation to de Duve's remark suggesting that evolution can work "only by exploiting randomness fully and leaving nothing to chance" (de Duve, '91: 213).

[6] See Keller (2005) for further discussion.

Of course, the question of just what role chance plays in evolution, and where it plays this role, is at the crux of many of the most heated debates about evolution. Neo-Darwinian theory, at least in its simplest version, confines chance to the genesis of point mutations, claiming that the rest of the work can be done – and in fact is done – by selection acting on the variation these mutations give rise to. Knowing what we now know both about the many other sources of inheritable variation, and about the complex regulatory dynamics governing the genesis and repair of point mutations, this view has come to seem naïve, but we have yet to see a satisfying alternative formulation that does justice to the interplay between chance and regulation in biological systems. In models of self-organizing systems, the role of chance is also confined to the micro level, albeit to molecular interactions or network construction, rather than to nucleotide change. Evolution by composition extends the role of chance to the macro level of modular interactions. Undoubtedly, chance operates on all these levels, and even more crucially, its effects on every level are filtered through and molded by selection. It is my belief that accounting for the evolution of function, design, and agency – the ultimate challenge – will almost certainly require a multi-level integration of this cumulative, time-dependent, interplay between selection and chance. More specifically, it seems to me that such an integration is needed if we are to include the effects of regulatory mechanisms on both the genesis and the impact of subsequent chance events, once these mechanisms have arrived on the scene.

Chapter 10
Self-Organization and Complexity in Evolutionary Theory, or, in this Life the Bread Always Falls Jammy Side Down

Michael Ruse

> *Let it be borne in mind in what an endless number of strange peculiarities our domestic productions, and, in a lesser degree, those under nature, vary; and how strong the hereditary tendency is. Under domestication, it may be truly said that the whole* organization *becomes in some degree plastic. Let it be borne in mind how infinitely complex and close-fitting are the mutual relations of all organic beings to each other and to their physical conditions of life. Can it, then, be thought improbable, seeing that variations useful to man have undoubtedly occurred, that other variations useful in some way to each being in the great and complex battle of life, should sometimes occur in the course of thousands of generations? If such do occur, can we doubt (remembering that many more individuals are born than can possibly survive) that individuals having any advantage, however slight, over others, would have the best chance of surviving and of procreating their kind? On the other hand we may feel sure that any variation in the least degree injurious would be rigidly destroyed. This preservation of favourable variations and the rejection of injurious variations, I call Natural Selection.*
>
> *(Charles Darwin,* Origin of Species, *80–81)*

I am an ardent Darwinian. I believe that life evolved and that the all-important mechanism was natural selection (Ruse 2006b). More organisms are born than can survive and reproduce, this leads to a struggle for existence, and only some succeed in passing on their bloodlines to future generations. Those that do succeed tend to have features different from the losers, it is these differences that make for success, and so over time there is heritable change (evolution) and this is in the direction of help-conferring characteristics, or adaptations. With John Maynard Smith and Richard Dawkins, I believe that selection produces this, the most interesting feature

M. Ruse (✉)
Professor, Department of Philosophy, Florida State University, FL, USA
e-mail: mruse@mailer.fsu.edu

A. Barberousse et al. (eds.), *Mapping the Future of Biology*, Boston Studies in the Philosophy of Science 266, DOI 10.1007/978-1-4020-9636-5_10,

of the organic world, that which most calls for explanation, namely the design-like nature of this world – its adaptedness or "organized complexity" (Ruse 2003). Natural selection (and only natural selection, including here sub-forms like sexual selection) speaks to this feature.

This means that I take organization very seriously indeed. So seriously in fact that I believe attempts to do away with – constrain, belittle, supplement – natural selection are fundamentally mistaken. Let me articulate and defend my position in two ways, first by showing how seriously I take organization and then by showing how little I regard alternatives.

10.1 What is Life?

Ask the most basic of biological questions: What is life? Most obviously, the answer is that life is some sort of substance. Rocks never had it. A dead cow had it and has now lost it. You my reader have it, and I have it. We have life, the cow had it and then it went, the rock was ever lifeless. But what kind of substance is this life? As an undergraduate, my wife majored in Animal and Poultry Science. In one never-to-be-forgotten course, a living pig was brought into the classroom, weighed, taken out and shot, and then the carcass returned and cut up, and the parts weighed and examined. The students dutifully noted down the size of the various bodily parts and compared them to estimates that they had made when the porker was still alive and squealing; but never were they expected to note down the weight of the living substance and to compare it with the weight of the dead substance, seeking (to use a term often employed in such discussions as these) the "life force." Such forces, if they exist, have no weight.

Not that this is necessarily a bar to thinking in terms of a substance. Most of us, in everyday life, follow philosophers like Plato and Descartes in thinking that the mind in some way is a force, and a substance – to each of us, it is as real as our hands or arms (and a lot more real than say our pancreas, that most of us could not locate or tell its function, if we were challenged). Yet no one thinks that the mind has weight. You do not lose ten pounds when you fall asleep. Perhaps we might say that the life force is a bit like a mind – not so much a conscious mind, for most of us do not think that trees have thoughts – but something akin to a mind that is working always and subconsciously. A kind of animating (or vegetating) force.

This was the position of the great Greek philosopher Aristotle, who was also a serious (and, as we now realize, very good) marine biologist. For living organisms he thought there was some kind of soul – not in the Christian sense (he wrote four hundred years before Jesus) or in a theological sense at all – but in the sense of something that makes for life and action. It is not just a question of the material out of which organisms are formed, for "we are inquiring not out of what the parts of an animal are made, but by what agency. Either it is something external which makes them, or else something existing in the seminal fluid and the semen; and this must either be soul or a part of soul, or something containing soul." He thought that it was something that exists "in the embryo itself..." – a kind of life force, making for vitality, for the very act of living and breathing and being. We may not see it directly, but just like the mind itself, it is there and functioning (Farley 1977; Fry 2000).

This idea proved extremely influential and it persisted in one way or another right down to the twentieth century. Indeed, it was around the beginning of this last century that – probably driven by dissatisfaction with the materialistic tendencies of modern biology – a number of people began to champion the view that a return to Aristotle is needed. Life itself – especially evolving life itself – demands more than a purely naturalistic approach. These people wanted to escape from a purely materialistic viewpoint – one that saw nature as driven by blind mechanisms – and they reverted to supposing that it is life forces that animate and push life forward. In Germany, leading this new "vitalism" was the embryologist Hans Driesch. He argued that we need to invoke something which he termed an "entelechies." In France, the leading figure was the philosopher Henri Bergson (1907) – partly Jewish but later to grow close to Catholicism.

Bergson was a deeply committed evolutionist. There was no compromise on gut positions. Nevertheless, Bergson judged that all then-existing theories of evolution – all then-existing purely mechanistic theories of evolution – were inadequate. For Bergson, as for so many others – Maynard Smith, Dawkins, Ruse – it was the complexity issue that was crucial. He thought that a mechanistic view destroys the holistic view that we need of the organism. "The real whole might well be, we conceive, an indivisible continuity. The systems we cut out within it would, properly speaking, not then be *parts* at all; they would *partial views* of the whole. And, with these partial views put end to end, you will not make even a beginning of the reconstruction of the whole,. . ." This decompositional approach spells the end of trying to understand the essence of life. "Analysis will undoubtedly resolve the process of organic creation into an ever-growing number of physicochemical phenomena, and chemists and physicists will have to do, of course, with nothing but these. But it does not follow that chemistry and physics will ever give us the key to life" (Bergson 1911, 32–3).

We need something that will give direction to evolution. We need something that will make for complexity and the like. Bergson located this creative power – something he christened the *élan vital* – in the life force or impetus possessed by all living things. "This impetus, sustained right along the lines of evolution among which it gets divided, is the fundamental cause of variations, at least of those that are regularly passed on, that accumulate and create new species" (pp. 92–3). The impetus, the *élan vital*, is like consciousness, deciding on the best path to be taken and then trying to walk along it. There is direction and the influence of the end, but not so much from the end itself as from the consciousness of the end. Defining life as "a tendency to act on inert matter," Bergson continued that: "The direction of this action is not predetermined; hence the unforeseeable variety of forms which life, in evolving, sows along its path. But this action always presents, to some extent, the character of contingency; it implies at least a rudiment of choice. Now a choice involves the anticipatory idea of several possible actions. Possibilities of action must therefore be marked out for the living being before the action itself" (pp. 101–2). Apparently, sight is just such a possibility of action, and that is why complex eyes have evolved several different times.

Bergson's ideas were highly influential. But for all this, we must admit how greatly out of tune with modern science was such thinking. The problem was not so much that the *élan vital* was unseen or directly unknowable. Twentieth-century science in particular is loaded with the unseen and directly unknowable. What are we to make of electrons, with their complementary qualities of particles and waves? What separated the *élan* from the electron was that ultimately whereas the latter notion is very useful – indispensable – the former notion is not. It gives the impression of explanatory power but it is not embedded in laws and cannot be used for prediction or unification or any of the other epistemic demands that one makes of the unseen entities of science. One can do just as much without the *élan* as one can do with it. The palaeontologist G. G. Simpson (1949) put his finger on things: "Granting, as any reasonable person must, that there is an important difference between life and non-life, you may, if you wish, call the different behaviour of matter in life 'vitalistic,' but this accomplished nothing and means nothing that was not already obvious. It is an example of the naming fallacy to call this an explanation or a contribution to evolutionary theory" (p. 125).

This was so before the days of the DNA molecule and the double helix, and is even more so after. The *élan vital* was of no help to biologists cracking the genetic code. Workers have found no place for the *élan vital* as they struggle to follow the development of the organism from the macromolecules of nucleic acid to the finished adult. And worse even than this. The *élan* seemed to commit one to some ideas deeply antithetical to modern science. From the time of the philosopher Francis Bacon, science has eschewed forces that impose direction on the course of events – "teleological" forces – and yet for all that Bergson claimed that he did not want to give consciousness to all living matter, that kind of teleology was precisely what he was introducing into science. And with this, he was attributing consciousness of organisms where such attribution just does not seem justified. If the trilobite, say, or the plant, did not have consciousness in some wise, then how could it choose to go in one direction rather than another?

10.2 Organization

If life is not a substance, what then is it? To use a much-hackneyed term, a paradigm shift was in the wind. Perhaps life is not really a substance at all. Although people agreed with Bergson when he seized on the issue of complexity as the issue of real significance, many thought that this might point the way to the solution rather than merely to the problem to be tackled. As the secrets of the cell and of heredity were being uncovered, the tremendous organization involved was what struck many. It is not so much that we have different materials – more and more, people were finding that the materials are basic chemical compounds – but the way they are put together. The way – and here again of course we go back to other aspects of Aristotle's thinking – things are put together and function. The difference between a clock that works and a clock that does not is not a question of "clock force," but that the one is put together properly and the other is not. The one is organized properly, and the other is not. Many interested in the nature of life, therefore, were drawn to dropping substance talk and taking up the language of organization.

One highly influential figure here was the great evolutionist of the early twentieth century, J. B. S. Haldane (1929). He argued explicitly that life is basically a question of organization. Take a play by Shakespeare. It is as much a verbal experience, as just something visual. "Be, food, if, love, music, of, on, play, the," is a list. "If music be the food of love, play on!" is the start of great literature. Of course the content matters. "It is important to know this, as it is important to know that life consists of chemical processes. But the arrangement of the words is even more important than the words themselves. And in the same way, life is a pattern of chemical processes" (Haldane 1949). Haldane stressed that these are processes of a particular kind, in particular processes that are able to keep themselves going and to replicate in some wise. "This pattern has special properties. It begets a similar pattern, as a flame does, but it regulates itself in a way that a flame does not except to a slight extent. And, of course, it has many other peculiarities. So when we have said that life is a pattern of chemical processes, we have said something true and important." What is true and important is that life is not a simple substance – a thing. It is rather a matter of the way in which elements are put together.

Of course, we cannot be talking about just any kind of organization. One might think of a crystal as organized, but it is not living. It has to be organization of a special kind, namely that which we associate with living things. And what is it that we associate with living things? Clearly reproduction and (if we are evolutionists) the ability to change and diverse. So the organization has to be put in terms of being something capable of self generating and of maintaining itself – even if we agree that a crystal can generate itself, it cannot maintain itself. Nor can it evolve. Of course, this is probably going to give us some borderline cases. Thunderstorms might seem to get close to qualifying. Anyone who lives in Florida and has been in the path of a tornado will not think this a joke. Closer to biology, is a virus something living or not? It perpetuates itself only parasitically on other organisms. Probably the right attitude is to say that borderline cases are what we get in the real world – especially the real world of evolution. Tornados, thank goodness, do not live and persist forever. So they are disqualified. Viruses are a judgment call, and therefore perhaps make the case rather than destroy it.

None of this is to trivialize the significance of life or to say that now its origin is at once explained – anything but. Rather it is to take unnecessary mystery out of the concept of life and recognize it for what it is rather than what it is not or might be. The dead pig weighs no less than the live pig, but its organization has been disturbed – its brain and its nerves and whatever has been messed up by the slaughterer's bullet and so it no longer works. It is no longer alive. Life is a matter of functioning and organization rather than substance. And with this shift in emphasis, the origin of life question became transformed – at once, more complex and yet more hopeful.

10.3 Is Natural Selection All-Powerful?

Leave the origin of life problem now. The significance of organization is now a given. But is it too much of a given? No biologist today is going to deny its significance. But are we overdoing the question of organization? In particular, has the Darwinian's enthusiasm for selection led to an over-evaluation of the organization of the living world? Put matters this way. If natural selection is all-powerful, then we would expect to find organization – which the Darwinian cashes out in terms of adaptation – everywhere, and that it is always works perfectly. It was Stephen Jay Gould who launched the strongest attack on this picture of life, in a well-known article, co-authored by Richard Lewontin: "The Spandrels of San Marco" (Gould and Lewontin 1979). They argued strongly that evolutionists – Darwinians particularly – assume far too readily that living nature is adaptive, that it is functional. They did not want to deny that the hand and the eye are adaptations, for clearly they are. But they felt that too often evolutionists slide into some kind of pan-adaptationism, thinking that every last organic feature has to be functional, the product of natural selection. Referring to the Leibnizian philosopher in Voltaire's *Candide*, they accused evolutionists of Panglossianism, thinking that these must be the best of all possible features in the best of all possible worlds. And to make the case complete, supposedly, evolutionists invent "just so" stories – thus named from Rudyard Kipling's fantasy stories – with natural selection scenarios leading to adaptation.

To counter this, Gould and Lewontin drew attention to the triangular decorative aspects of the tops of pillars in medieval churches. They argued that although such "spandrels" seem adaptive – areas for creative outpourings – in fact they are just by-products of the builders' methods of keeping the roof in place. "The design is so elaborate, harmonious, and purposeful that we are tempted to view it as the starting point of any analysis, as the cause in some sense of the surrounding architecture." This, however, is to get things precisely backwards. "The system begins with an architectural constraint: the necessary four spandrels and their tapering triangular form. They provide a space in which the mosaicist worked; they set the quadripartite symmetry of the dome above" (Gould and Lewontin 1979, 148). Perhaps, argued the two men, we have a similar situation in the living world. Much that we think adaptive is merely a spandrel, and such things as constraints on development prevent anything like an optimally designed world. Perhaps things are much more random and haphazard – non-functional, non-organized – than the Darwinian thinks possible.

Now, what is to be said by the Darwinian in response to this charge? Simply this: Whoever doubted the point that Gould and Lewontin are making? It has always been recognized by evolutionists – certainly from the *Origin of Species* on – that however common or ubiquitous adaptation may be, it is only part of the story. The living world is not – cannot be – totally and completely adaptive. In fact, this is one of the strongest points against the God of the natural theologians. There is far too much wrong with the world – too many instances of malfunction – to think that a designer has been directly involved with making organisms. Perhaps Gould and Lewontin are right in thinking that, too often, adaptation has been assumed when it clearly does not exist. Although, the Darwinian would point out, it is often only by assuming adaptation that one can expect to bully through to the answer.

More significantly, Gould and Lewontin are reinventing the wheel if they think that they are drawing attention to some new or hitherto-neglected problem or issue. No Darwinian would deny that sometimes, to take an example that Gould and Lewontin endorse, you just get things that do not fit or that lag behind in some way, and that selection fails to correct. The ornithologist Nicholas Davies (Davies and Brooke 1988) studies European cuckoos and the birds that they parasitize. He has shown that the cuckoos have many adaptations designed to make them successful in their aims at dropping their eggs in the nests of others – for instance, whereas most birds like to linger over the laying process, cuckoos are in and out in a flash. One area where there is extreme competition between cuckoos and hosts – what evolutionists call an "arms race" – focuses on egg color. Robins and most other birds are very good at spotting alien eggs, and cuckoos are very good at camouflaging their eggs to look like those of the hosts. The exception is the dunnocks, hedge sparrows. They are parasitized by cuckoos, but lack the ability to discriminate between their own eggs and those of others, and in turn the cuckoos make no effort to disguise their eggs. There is a clear adaptive breakdown here, for the dunnocks lose out.

The answer apparently is that dunnocks have been only recently parasitized and have not yet built up adaptations to the cuckoos. A clever experiment, using model eggs of various kinds, provided support for this belief. "The cuckoo breeds from Western Europe to Japan but not in Iceland, where it is only a rare vagrant and has never been known to breed. Iceland does, however, have isolated populations of meadow pipits and white wagtails (of which the pied wagtail is a subspecies). We therefore took our model eggs to Iceland. The Icelandic populations bred at low densities and we had to work very hard to find nests but the results were exciting. Both the pipits and wagtails showed much less discrimination against eggs unlike their own than did members of the parasitized populations of these two species in Britain" (Davies 1992, 229–30). Clearly here the power of selection has not proven all-powerful.

Sometimes we get adaptive failure – or at least less than adaptive excellence – because, as Darwin always stressed, one is not setting down a blueprint for excellence and then producing it. One is simply trying to do better than others. As the old joke has it: it is not necessary to be super fast to escape the bear in the wood, merely faster than the chap next to you. There is no need of adaptive excellence in some absolute sense. Factors like this were surely important in the early years of the Cambrian explosion. It was not that the new complex forms were super-efficient. It was rather that they did better than others and had no more-complex rivals to wipe them out. Evolution is a bit like trying to cross a desert in an old car. If it breaks down, you have to find some way to get things going with what you have – rather like the astronauts in Apollo Thirteen – rather than going back to the garage or the drawing board and starting again. This means that life – like the old car – will reflect the contingencies of the situation. Thanks to selection, life may make moves which work but which in the fullness of time no one would think ideal or even the best. Take human birth. No one would say that this is a model of adaptive excellence. It is a matter of compromise. You have animals that are bipedal and you have animals that put a premium on intelligence. You want the babies to have brains as big as possible by birth. And so, what you get are difficult and dangerous births – not as difficult and dangerous as to rule out all successful birthing, but threatening to both mothers and babies, nevertheless.

Sometimes, we just get stuck with history or with the reflection of history. Take homologies, notably the vertebrate case highlighted by Darwin: the bones of the forelimb. The hand of human, the leg of horse, the flipper of dolphin, the wing of bird (and differently of bat) are all built on the same pattern, the same archetype. There is no direct functional reason for any of this. It is part of history. But, complains the critic, is not this precisely the sort of factor that means Darwinism is exhausted?

Yet, truly, does this render selection impotent or irrelevant? It does stress that selection does not start every new experiment from scratch, but cobbles together the working parts from what exists originally. But this does not now mean that selection has no power or that it does not do its job. Selection took what there was and molded it to different ends. The resulting adaptations could always be very good. No one denies that the human forelimb and the horse's leg work quite brilliantly. And selection deserves the credit. Not that this is to deny that, more dramatically, sometimes nature just goes the wrong way, and we show a history that is not very helpful – adaptation that is really not as good as it might be. Consider the human, male, urogenital system, where, thanks to our evolutionary history, the sperm duct is rather like a garden hose which takes an unneeded loop around a distant tree, on its way from tap to nearby flower bed. The duct got itself hung over the urethra, and so instead of going directly from testes to penis, it meanders around the body before it gets on with the job (Williams 1992a). Hardly a triumph of plumbing, although note that there is nothing surprising about any of this, and certainly nothing that threatens Darwinism or adaptationism as such. No one is saying that the urogenital system is not adaptive or that you would be better off without it. The point simply is that you must think in a relativistic sense – is it better than the competitors? – rather than an absolutist sense – is it the best that could ever be?

10.4 Constraints

Continuing the defence of the significance of organization, let us turn now to a recently much-discussed topic, namely that of constraint. Gould, and other critics of the ubiquity of adaptation, argue that often – too often – selection cannot work and adaptation fails because organisms are constrained in one way or another. "Organisms are capable of an enormous range of adaptive responses to environmental challenge. One factor influencing the pathway actually taken is the relative ease of achieving the available alternatives. By biasing the likelihood of entering onto one pathway rather than another, a developmental constraint can affect the evolutionary outcome even when it does not strictly preclude an alternative outcome"

(Maynard Smith et al. 1985, 269). Thus defined, no one would ever deny that there are constraints and that they can be significant. Homology – skeletal and molecular – concedes this. Supposedly we have historical constraints. Because of organisms' history – where they were located at a certain point in their history – constrains the subsequent moves and the power of selection. The vertebrate forelimb cannot be of any form it wants. It must be a modification of the proto limb of the past. Likewise with the genes for development, if truly there was no other way in which things could be done.

But what about developmental homology? Leading students of the new field of evolutionary development – "evo devo" – think that here the non-adaptive argument bites more deeply.

> The homologies of process within morphogenetic fields provide some of the best evidence for evolution – just as skeletal and organ homologies did earlier. Thus, the evidence for evolution is better than ever. The role of natural selection in evolution, however, is seen to play less an important role. It is merely a filter for unsuccessful morphologies generated by development. Population genetics is destined to change if it is not to become as irrelevant to evolution as Newtonian mechanics is to contemporary physics (Gilbert et al. 1996, 368).

Let us dig more deeply here, using the critics' own words as our subject matter. Consider biological constraints. The developmental morphologist Rudolf Raff (1996) raises the issue of genome size. "Having a large genome has consequences outside of the properties of the genome per se. Larger genomes result in larger cells. Because cells containing large genomes replicate their DNA more slowly that cells with a lower DNA content, large genomes might constrain organismal growth rates. Cell size will also determine the cell surface-to-volume ratio, which can affect metabolic rates" (p. 304). Salamanders often have large genome sizes and thus are good organisms on which to test hypotheses about constraints. And there does seem to be some evidence of their operation. "Roth and co-workers have observed that in both frogs and salamanders, larger genome size results in larger cells. In turn, larger cells result in a simplification of brain morphology. Thus, quite independently of the demands of function, internal features such as genome size can affect the morphology and organization of complex animals. Plethodontid salamanders share the basis vertebrate nervous system and brain, but they have very little space in their small skulls and spinal chords" (p. 305, referring to Roth et al. 1994).

Yet, Raff has to admit that if there are constraints, they cannot be that overwhelmingly significant. The salamanders can still do some pretty remarkable things. They certainly do not come across as organisms desperately functionally constrained. "These salamanders occupy a variety of caverniculous, aquatic, terrestrial, and arboreal habitats. They possess a full range of sense organs, and most remarkably, a spectacular insect-catching mechanism consisting of a projectile tongue that can reach out in ten milliseconds to half the animal's trunk length (snout to vent is the way herpetologists express it)." They have pretty good depth perception too. Paradoxically, in fact, their slow metabolic rate brought on by large genome size may even be of adaptive advantage. "Plethodontids are sluggish, and the low metabolic rates introduced by large cell volume may be advantageous to sit-patiently-and-wait hunters that can afford long fasts. Vision at a distance is reduced to two handbreadths, but since these animals are ambush hunters that strike at short range, that probably doesn't affect their efficiency much" (p. 306). All things considered, nothing here need keep the ardent selectionist from a good night's sleep. And if insomnia does threaten, then there is the additional bromide that apparently, if need be, the salamanders can start to bring down their genome sizes.

Turning from the biological to the physical, there are many constraints influencing organisms and setting limits on what organisms can and cannot do, and where and when they should invest their energies. Take one of the most basic of all facts: size and consequent weight go up rapidly, according to the cube power of length or height. Consider two identically shaped mammals, and suppose one to be twice the height of the other. The taller mammal is going to be eight times as heavy as the shorter mammal. From a structural perspective, it has eight times the weight problem. Do not ask why you cannot build elephants as agile as cats. They are a physical impossibility. They would need far more support, and this would mean that they would need bigger and stockier (non-cat-like) bones (Vogel 1988).

Another interesting calculation concerns what has irreverently been called the "Jesus number." What are the constraints on walking on water? A fairly simple formula governs the activity. Pushing up is the surface tension, γ, times the perimeter of the feet or area that is touching, l. Pushing down is gravity, which is a function of the mass, m, times the gravitational attraction, g ($F = mg$), or restating in terms of density, ρ, times the volume, which is a function of l cubed. In other words: $\mathrm{Je} = \gamma l/\rho l^3 g = \gamma/\rho l^2 g$. Since the surface tension, the density, and the gravitational constant remain the same, this means that the ability to walk on water is essentially a function of the perimeter squared. In other words, the smaller you are the better off you are, and conversely the bigger you are the more likely you are to sink. This is no problem for insects, especially given that they have six legs and so have a long perimeter compared to the body size. Humans however are another matter. "What would be the maxim weight of a human who could walk on water? My size-9 sandals have a perimeter of 0.62 meters each; that length times the surface tension of water gives 0.045 newtons of force, or 4.6 grams (less than half an ounce) of weight – 9.2 grams to stand (two feet in contact) or half that to walk. The theological implications are beyond the scope of the present book" (Vogel 1988, 100).

Physical constraints are important, although there are times when it is not obvious that such constraints should really be called "constraints." John Maynard Smith and his co-authors have explored in depth the example of the coiling of shells in such organisms as mollusks and brachiopods. The coiling itself is fairly readily reduced to a simple logarithmic equation and it is possible to draw a plain that maps the coiling as a function of the vital causal factors, particularly the rate of coiling and the size of the generating curve. Given such a map, one feature stands right out for comment: whereas for most shells the coils touch all the way from the centre to the perimeter, some such shells coil without touching. There is a gap between the coils. Now, map the actual shells of a group of organisms, the genera of extinct ammonoids (cephalopod mollusks). The isomorphism between the theoretical and the actual is outstanding.

> [N]early all ammonoids fall on the left side of the curved line and thus display overlap between successive whorls. This is clearly a constraint in the evolution of the group but what kind of constraint? In this particular case, the answer is apparently straightforward (Raup 1967). Evolving lineages can and occasionally do cross the line so there is no reason to believe that open coiling violates any strict genetic or developmental constraint. Rather, the reason for not crossing the line appears to be biomechanical. Other things being equal, an open coiled shell is much weaker than its involute counterpart. Also, open coiling requires more shell material because the animal cannot use the outer surface of the previous whorl as the inner surface of the new whorl. (Maynard Smith et al. 1985, 280)

Even the exceptions prove the point. The shell of the living pelagic cephalopod Spirula has a shell that coils but does not touch. Exceptionally, this organism carries the shell internally, using it for buoyancy. There is no need for strength. Maynard Smith and co-authors conclude "that the constraint against open coiling is an adaptive one brought about by simple directional selection." A conclusion which surely brings us full circle, for if constraints can be *adaptive*, brought on by selection, the whole argument against the ubiquity of adaptation is knocked sideways somewhat. Organized complexity rules supreme.

10.5 Order for free

Let us come to the climax of this paper. Perhaps Gould's fault was that he was too timid rather than too daring. Perhaps the real issue is not complexity but how it comes about. Perhaps the Darwinian is right to stress the problem of organized complexity, but wrong to stress the importance of natural selection. Perhaps the anti-Darwinian needs a much more radical attack on the need for natural selection or its place as the central mechanism of evolutionary change. Lamarckism will never replace natural selection, and neither will traditional, undirected variations. Yet, have we eliminated the possibility that the regular, unguided laws of physics can do all that is needed – that they can entirely replace or significantly supplement selection? The focus now is on complexity, whether with Darwinians you think that this is adaptive complexity, or whether you think that it is just complexity and leave it at that.

Almost from the time of Darwin on – certainly through the last century – there have been those who think that natural selection is unneeded, a red herring (and a rather smelly one at that), and that the laws of physics and chemistry can do the job unaided. The leader of this school was the Scottish morphologist D'Arcy Wentworth Thompson. His work, *On Growth and Form*, first published in 1916, is the founding bible. Thompson was one who took things to the extreme, denying that the organic world shows much design – he certainly had no time for the search for adaptation, thinking that this just holds biology back. "To seek not for ends but for antecedents is the way of the physicist, who finds 'causes' in what he has learned to recognize as fundamental properties, or inseparable concomitants, or unchanging laws, of matter and of energy. In Aristotle's parable, the house is there that men may live in it; but it is also there because the builders have laid one stone upon another" (Thompson 1948, 6). Continuing: "Cell and tissue, shell and bone, leaf and flower, are so many portions of matter, and it is in obedience to the laws of physics that their particles have been moved, molded and conformed.... Their problems of form are in the first instance mathematical problems, their problems of growth are essentially physical problems, and the morphologist is, *ipso facto*, a student of physical science" (p. 10). Thus: "We want to see how, in some cases at least, the forms of living things, and of the parts of living things, can be explained by physical considerations, and to realize that in general no organic forms exist save such as are in conformity with physical and mathematical laws" (p. 15).

Thompson's approach to evolutionary questions was, to put it gently, somewhat fuzzy. Probably, despite his rhetoric, he was not entirely against adaptation as such. However, he surely wanted to downplay its existence and significance – at best, it was a corollary of the development of form, as governed by the principles of physics and chemistry. Complexity exists and complexity is important, whether or not you tie it in with adaptation. A paradigmatic case of nature's complexity that spurred Thompson to eloquence was the shape of the jellyfish. He saw this as a straight consequence of the physics of drops of liquid of one density falling in a liquid of a different, somewhat lower density. The patterns of the falling liquid are precisely the patterns of the organic, liquid-dwelling jellyfish. "The living medusa has a geometrical symmetry as marked and regular as to suggest a physical or mechanical element in the little creature's growth and construction.... It is hard indeed to say how much or little all these analogies imply. But they indicate, at the very least, how certain simple organic forms might be naturally assumed by one fluid mass within another, when gravity, surface tension and fluid friction play their part, under balanced conditions of temperature, density and chemical composition" (pp. 396–8).

To be fair to Gould's credentials as a Darwin basher, some thirty years ago he did write an appreciative essay of Thompson's labours. But it has been others who have done the spade work, notably the Canadian-born, English-dwelling morphologist Brian Goodwin (2001) and (in America) the medically trained, theoretical biologist Stuart Kauffman. The latter particularly, with the driven dedication of a prince seeking the Sleeping Beauty, has cut through the thickets of "self organization" or (as Kauffman cleverly calls it) "order for free," finding physics lying there waiting for his kiss to wake it from its slumbers. "The tapestry of life is richer than we have imagined. It is a tapestry with threads of accidental gold, mined quixotically by the random whimsy of quantum events acting on bits of nucleotides and crafted by selection sifting. But the tapestry has an overall design, architecture, a woven cadence and rhythm that reflect underlying law – principles of self organization" (Kauffman 1995, 185). Seizing on another example instanced by Thompson, today's formalists (as we may call them) invite us to consider the phenomenon of phyllotaxis, the pattern of clockwise and anticlockwise spirals shown by many plants, where identical elements are packed together. Sunflowers, pinecones, even the lowly cauliflower – they all exhibit this intricate pattern. There is no chance here. The pattern, phyllotaxis, is produced by the leaves or analogous plant parts appearing at the centre (the "growing apex") and then, as it were, being pushed outwards (Mitchison 1977). The appearing leaves or parts follow a twisted path ("known as the genetic spiral") and, if growth is constant, then the angle between successive leaves or parts is constant.

It is not the genetic spiral that catches the eye; rather it is the criss-crossing diagonal spirals (known technically as "parastichies"). Pre-evolutionists realized that one can express phyllotaxis in mathematical form by means of a formula discovered by the thirteenth-century Italian mathematician, Leonardo Fibonacci. Searching for a way of calculating the growth of the offspring of a pair of rabbits, he arrived at the series where any member is formed by adding together the previous two members of the series, starting with zero and one. Thus one has 0, 1, 2, 3, 5, 8, 13, ..., or more generally, $n_j = n_{j-1} + n_{j-2}$. It turns out that, for any particular species of plant, the numbers of parastichies, one set clockwise and one set counter-clockwise, are always related by being consecutive numbers of the Fibonacci series. In the stylized picture given in the diagram, the example is of a 3, 5 phyllotaxis. As can also be seen from the diagram, another way of calculating the measure is by using the order of production of the "contact" leaves on the same spiral. This is not a measure based on the order of production of the leaves but on the pattern itself, and refers to those leaves, along shared paratischies, that will be touching. Using examples furnished by Asa Gray in the sixth edition of his textbook, *Structural Botany* (1881), the American larch produces a cone that is 2,3, holly is 3,5, and the cone of *Pinus strobus* is 5,8.

Darwinians have jumped on this phenomenon with joy. In the early 1870s, the American pragmatist, Chauncey Wright, argued that this kind of arrangement gives the best way of exposing each leaf to the light, without undue overlap from its fellows. He argued that the differences between the various phyllotactic arrangements are so minute as not really to matter that much. "To realize simply and purely the property of the most thorough distribution, the most complete exposure to flight and air around the stem, and the most ample elbow-room, or space for expansion in the bud, is to realize a property that exists separately only in abstraction, like a line without breadth" (Wright quoted in Gray 1881, 125). People like D'Arcy Thompson and Brian Goodwin will have none of this. Thompson listed one objection after another. He thought that the differences between the arrangements is indeed significant, the teleological intent is something which "cannot commend itself to a plain student of physical science," there are all sorts of other ratios that would do the job as well, the plant could have taken other and better paths to exposing the leaves to sunlight, and much more. "We come then without more ado to the conclusion that while the Fibonacci series stares us in the face in the fir-cone, it does so for mathematical reasons; and its supposed usefulness, and the hypothesis of its introduction into plant-structure through natural selection, are matters which deserve no place in the plain study of botanical phenomena" (Thompson 1948, 953).

My sense is that today's complexity theorists are more favourably inclined to adaptation than Thompson. Where as he would see phyllotaxis as complex and leave matters at that, someone like Goodwin would see it as complex and supporting adaptation. But even if – especially if – you interpret things in this second way, it will be no surprise to learn that Darwinians are not convinced. They argue that the complexity theorists overlook the "obvious possibility" that "natural selection may universally favor close packing by phyllotaxis over alternative arrangements" (Reeve and Sherman 1993, 21). It is not to be denied that phyllotaxis is produced by development going down certain, fixed channels. How else would it occur? But form does not preclude function. Even if the patterns as such are fixed, nothing too much matters so long as other plant features can be varied in order to maximize leaf exposure efficiency. Which seems to be the case: "Computer simulations indicate that phyllotaxy can influence the quantity of light intercepted by leaf surfaces. Model plants constructed with equal total leaf area and number differ significantly in flux, even when leaf-divergence angles are very similar.... Nonetheless, computer simulations indicate that a variety of morphological features can be varied, either individually or in concert, to compensate for the negative aspects of leaf crowding resulting from 'inefficient' phyllotactic patterns. Internodal distance and the deflection (tilt) angle of leaves can be adjusted in simulations with different phyllotactic patterns to achieve equivalent light-interception capacities" (Niklas 1988, 566). The basic point is that however it is caused, complexity does not just happen. For its systematic appearance and persistence – as in the case of phyllotaxy – there has to be a tie-in with selection. It is not a question of form or function. Rather, it is function underlaid by form. Physics may produce the complexity. Natural selection cherished and refines the complexity, to its own needs.

Ultimately, of course, it is a metaphysical issue at stake here, not a scientific one. The fact is that the living world does seem complex and D'Arcy Thompson notwithstanding, as adaptive – as if designed. For the Darwinian, blind law will not account for this. Blind law leads to mess, to decay, to (in the language of the physicists) increase in entropy. Murphy's Law holds: "If something can go wrong, it will," or the alternative version, "Bread always falls jammy side down." There must be a reason for the design-like nature of the world, and physics and chemistry just will not do. That is all there is to be said.

10.6 Conclusion

Complexity is of vital interest to the Darwinian. Natural selection is the cause. Attempts to replace selection or to deny its necessity are simply mistaken.

References

Adami, C., C. Ofria and Collier, T. C. 2000. Evolution of biological complexity. *Proceedings of the National Academy of Science USA* 97: 4463–4468.

Alberch, P. 1982. Developmental constraints in evolutionary processes. In: Bonner, J. T. (ed.) *Evolution and Development: Report of the Dahlem Workshop on Evolution and Development*. New York: Springer-Verlag, pp. 313–332.

Alexander, R. D. 1979. *Darwinism and human affairs*. Seattle, WA: University of Washington Press.

Allen, C., Bekoff, M., and Lauder, G. (eds.) 1997. *Nature's purposes: analyses of function and design in biology*. Cambridge: MIT Press.

Amundson, R. 2005. *The changing role of the embryo in evolutionary thought*. Cambridge: Cambridge University Press.

Arms, K. and Camp, P. S. 1995. *Biology*. (4th edn.). Fort Worth: Saunders College Publishing.

Arthur, W. 2004. *Biased embryos and evolution*. Cambridge: Cambridge University Press.

Ashby, W. Ross 1956. *An introduction to cybernetics*. New York: Chapman and Hall.

Avital, E. and Jablonka, E. 2000. *Animal traditions: Behavioural inheritance in evolution*. Cambridge: Cambridge University Press.

Azevedo, R. B. R. et al. 2006. Sexual reproduction selects for robustness and negative epistasis in artificial gene networks. *Nature* 440: 87–90.

Barkow, J. H. 1991. Precis of *Darwin, sex and status: Biological approaches to mind and culture. Behavioral and Brain Sciences* 14: 295–334.

Barkow, J. H., Cosmides, L., and Tooby, J. (eds.) 1992. *The adapted mind: Evolutionary psychology and the generation of culture*. Oxford: Oxford University Press.

Bedau, M. A. 1991. Can biological teleology be naturalized? *The Journal of Philosophy* 88: 647–655.

Bedau, M. A. 1992. Where's the good in teleology? *Philosophy and Phenomenological Research* 52: 781–806. Reprinted in Colin Allen, Marc Bekoff, George Lauder (eds.) 1997. *Nature's purposes: analyses of function and design in biology*, pp. 261–291. Cambridge: MIT Press.

Bedau, M. A. 1997. Weak emergence. In Tomberlin, J. (ed.) *Philosophical Perspectives: Mind, Causation, and World*, vol. 11, Oxford: Blackwell Publishers, pp. 375–399.

Bedau, M. A. 1998. Philosophical content and method of artificial life. In: Bynam, T. W. and Moor, J. H. (eds.) *The Digital Phoenix: How Computers are Changing Philosophy*. Oxford: Basil Blackwell, pp. 135–152.

Bedau, M. A. 2003a. Artificial life: organization, adaptation, and complexity from the bottom up. *Trends in Cognitive Science* 7: 505–512.

Bedau, M. A. 2003b. Downward causation and autonomy in weak emergence. *Principia* 6: 5–50. Reprinted in Bedau, M. A. and Humphreys, P. (eds.) 2007. *Emergence: Contemporary Readings in Philosophy and Science*. Cambridge: MIT Press.

Bedau, M. A. and Brown, C. T. 1999. Visualizing evolutionary activity of genotypes. *Artificial Life* 5: 17–35.

Bedau, M. A., McCaskill, J. S., Packard, N. H., Rasmussen, S., Adami, C., Green, D. G., Ikegami, T., Kaneko, K. and Ray, T. S. 2000. Open problems in artificial life. *Artificial Life* 6: 363–376.

Behe, M. J. 1999. Review *God After Darwin: A Theology of Evolution* by John Haught. *Metaviews*, Dec. 4.

Beldade, P., Koops, K., and Brakefield, P. M. 2002. Modularity, individuality, and evo-devo in butterfly wings. *Proceedings of the National Academy of Sciences, USA* 99: 14262–14267.

Beldade, P., Brakefield, P. M., and Long, A. D. 2005. Generating phenotypic variation: prospects from "evo-devo" research on *Bicyclus anynana* wing patterns. *Evolution and Development* 7:101–107.

Benirschke, K. 2006. Comparative Placentation: http://www.medicine.ucsd.edu/cpa/

Bergson, H. 1907. *L'évolution créatrice*. Paris: Alcan. *Creative evolution* (trans.), 1911, New York: Holt.

Bolker, J. 1995. Model systems in developmental biology. *Bioessays* 17: 451–455.

Bolker, J. A. and Raff, R. A. 1997. Beyond worms, flies, and mice: It's time to widen the scope of developmental biology. *Journal of NIH Research* 9: 35–39.

Berkeley, G. 1710/1965. *Principles of human knowledge*. Reprinted in Armstrong, D. (ed.) *Berkeley's Philosophical Writings*. New York: Collier.

Berlekamp, E. R., Conway, J. H., and Guy, R. K. 1982. *Winning ways for your mathematical plays*. vol. 2. New York: Academic Press.

Bennett, C. H. 1987. Demons, engines and the second law. *Scientific American* 275: 88–96.

Bock, W. J. 1980. The definition and recognition of biological adaptation. *American Zoologist* 20: 217–227.

Boniolo, G. 2003a. Biology without information. *History and Philosophy of the Life Sciences* 25: 257–275.

Boniolo, G. 2003b. Che cos'è il caso in biologia? Riflessioni a partire da J. Monod. *Il caso e la necessita. Atti dei Convegni Lincei* 185: 25–44.

Boogerd, F. C., Bruggeman, F. J., Richardson, R. C., Stephan, A., and Westerhoff, H. V. 2005. Emergence and its place in nature: a case study of biochemical networks. *Synthese* 145: 131–164.

Boogert, N. J., Paterson, D. M., and Laland, K. N. 2006. The implications of niche construction and ecosystem engineering for conservation biology. *BioScience* 56: 570–578.

Borel, E. 1962/1943. *Probability and Life*. New York: Dover Publications.

Brakefield, P. M., Gates, J., Keys, D., Kesbeke, F., Wijngaarden, P. J., Monteiro, A., French, V., and Carroll, S. B. 1996. Development, plasticity and evolution of butterfly eyespot patterns. *Nature* 384: 236–242.

Bresler, S. E. 1971. *Introduction to molecular biology*. New York: Academic Press.

Britten, R. J. and Kohne, D. E. 1968. Repeated Sequences in DNA. *Science* 161(3841): 529–540.

Burian, R. M. 1997. On conflicts between genetic and developmental viewpoints – and their attempted resolution in molecular biology. In: Dalla Chiara, M. L. (ed.) *Structures and Norms in Science*. Dordrecht: Kluwer, pp. 243–264.

Buss, L. 1987. *The evolution of individuality*. Princeton: Princeton University Press.

Caburet, S., Cocquet, J., Vaiman, D., and Veita, R. A. 2005. Coding repeats and evolutionary 'agility'. *BioEssays* 27: 581–587.

Calcott, B. 2006. Transitions in Biological Complexity. *Philosophy*. Canberra, Australian National University. Ph.D.

Campbell, J. 1982. *Grammatical man*. New York: Simon and Schuster.

Cartwright, N. 1989. *Nature's capacities and their measurement*. Oxford: Oxford University Press.

Cavalli-Sforza, L. L. and Feldman, M. W. 1981. *Cultural transmission and evolution: a quantitative approach*. Princeton: Princeton University Press.

Chadarevian (de), S. and Kamminga, H. (eds.) 1998. *Molecularizing biology and medicine: new practices and alliances 1910s–1970s*. London: Taylor and Francis.

Chemla, Y. R., Aathavan, K., Michaelis, J., Grimes, S., Jardine, P. J. Dwight, Anderson, L., and Bustamante, C. 2005. Mechanism of force generation of a viral DNA packaging motor. *Cell* 122: 683–692.

Colburn, T., Dumanoski, D., and Myers, J. P. 1996. *Our stolen future*. New York: Dutton.
Cook, C. E., Jimenez, E., Akam, M., and Salo, E. 2004. The Hox gene complement of acoel flatworms, a basal bilaterian clade. *Evolution and Development* 6: 154–163.
Copeland, D. C. 2000. The constructivist challenge to structural realism. *International Security* 25: 187–212.
Corbo, J. C., Di Gregorio, A., and Levine, M. 2001. The ascidian as a model organism in developmental and evolutionary biology. *Cell* 106: 535–538.
Crick, F. 1966. *Of molecules and men*. Seattle: University of Washington Press.
Crick, 1981. *Life itself its origin and nature*. New York: Simon and Schuster. Integrative and Comparative Biology 2006 46(5): 569–576.
Daly, M. and Wilson, M. 1988. *Homicide*. New York: Aldine de Gruyter.
Danks, H. V. 1987. Insect dormancy: an ecological perspective. *Biological Survey of Canada*. Ottawa: National Museum of Natural Sciences.
D'Arcy Thompson, W. 1948. *On growth and form, second edition*. Cambridge: Cambridge University Press.
Darling, J. A, Reitzel, A. R., Burton, P. M., Mazza, M. E., Ryan, J. F., Sullivan, J. C., and Finnerty, J. R. 2005. Rising starlet: the starlet anemone, *Nematostella vectensis. Bioessays* 27: 211–221.
Darwin, C. 1859/1964. *On the origin of species*. (Facsimile of first edition.) Cambridge MA: Harvard University Press.
Darwin, C. 1881. *The formation of vegetable mold through the action of worms, with observations on their habits*. London: Murray.
Darwin, C. 1883. *The variation of animals and plants under domestication*. (2nd edn.). New York: D. Appleton & Co., vol. 2, p. 282.
Davidson, E. 1995. *The regulatory genome; gene regulatory network in development and evolution*. Amsterdam, Boston: Elsevier/Academic Press.
Davidson, E. 2001. *Genomic regulatory systems*. New York: Academic Press.
Davies, N. B. 1992. *Dunnock behaviour and social evolution*. Oxford: Oxford University Press.
Davies, N. B. and Brooke, M. de L. 1988. Cuckoos versus reed warblers: Adaptations and counter-adaptations. *Animal Behaviour* 36: 262–84.
Davis, R. H. 2004. The age of model organisms. *Nature Review Genetics* 5: 69–76.
Dawkins, R. 1976. *The selfish gene*. Oxford, England: Oxford University Press.
Dawkins, R. 1982. *The extended phenotype*. Oxford: Freeman.
Dawkins, R. 1986. *The blind watchmaker*. New York: W. W. Norton.
Dawkins, R. 1995. *River out of Eden: A Darwinian view of life*. New York: BasicBooks.
Dawkins, R. 2001. Reply to Phillip Johnson. In: Pennock, R. T. (ed.) *Intelligent design creationism and its critics: Philosophical, theological, and scientific perspectives*. Cambridge, MA: Bradford/MIT Press, pp. 549–550.
Dawkins, R. 2004. *The Ancestor's tale: a pilgrimage to the dawn of life*. London: Weidenfeld and Nicholson.
Dawkins, R. 2005. Introduction: The Illusion of Design. *Natural History*. http://findarticles.com/p/articles/mi_m1134/is_9_114/ai_n15855364
DeAngelis, D. L. 1992. *Dynamics of nutrient cycling and food webs*. New York: Chapman and Hall.
de Duve, Christian, 1991. *Blueprint for a cell*. Burlington, NC: Neil Patterson Publishers.
Dedel, E. and Alon, U. 2005. Optimality and evolutionary tuning of the expression level of a protein. *Nature* 436: 588–592.
Delsuc, F., Brinkmann, H., Chourrout, D., Philippe, H. 2006. Tunicates and not cephalochordates are the closest living relatives of vertebrates. *Nature* 439: 965–968.
Dembski, W. 1999. Signs of intelligence: A primer on the discernment of intelligent design. *Touchstone Magazine* 12(4): 76–84.
Dembski, W. 2001. Intelligent design as a theory of information. In: Pennock, R. T. (ed.) *Intelligent design creationism and its critics: Philosophical, theological, and scientific perspectives*. Cambridge, MA: Bradford/MIT Press, pp. 553–573.

Dennett, D. C. 1995. *Darwin's dangerous idea: evolution and the meanings of life*. New York: Simon and Schuster.

Depew, D. J. and Weber, B. H. 1995. *Darwinism evolving: systems and the genealogy of natural selection*. Cambridge, MA: MIT Press.

Depew, D. J. and Weber, B. H. 2001. Developmental systems, Darwinian evolution, and the unity of science. In: Oyama, S., Griffiths, P. E., and Gray, R. D. (eds.) *Cycles of Contingency: Development System and Evolution*. Cambridge, MA: MIT Press, pp. 239–254.

Dewey, J. 1929/1958. *Experience and nature*. (revised edition). New York: Dover Reprints.

Dietrich, W. E. and Perron, J. T. 2006. The search for a topographic signature of life. *Nature*. 439. 411–418.

Donahue, K. 2005. Niche construction throough phonological plasticity: life history dynamics and ecological consequences. *New Phytologist* 166. 83–92.

Durham, W. H. 1991. *Coevolution: genes, culture and human diversity*. Paloanto, CA: Stanford University Press.

Dyson, F. 2005. The Darwinian Interlude. *Technology Review*, March.

Edmonds, B. 1996. What is complexity? In: Heylighen, F. Bollen, J., and Riegler, A. (eds.) *The evolution of complexity*. Dordrecht: Kluwer, pp. 1–16.

Eldredge, N. 2003. The Sloshing Bucket: How The Physical Realm Controls Evolution. In: Crutchfield, J. P. and Schuster, P. (eds.) *Evolutionary dynamics: exploring the interplay of selection, accident, neutrality and function*. Oxford: Oxford University Press.

Eldredge, N. and Gould, S. J. 1972. Punctuated equilibria: an alternative to phyletic gradualism. In: Schopf, T. J. M. (ed.) *Models in paleobiology*. San Francisco: Freeman, Cooper & Co., pp. 82–115.

Emlen, D. J. 1994. Environmental control of horn length dimorphism in the beetle *Onthophagus acuminatus* (Coleoptera: Scarabaeidae). *Proceedings of the Royal Society London* B 256: 131–136.

Emlen, D. J. 1996. Artificial selection on horn length-body size allometry in the horned beetle *Onthophagus acuminatus* (Coleoptera: Scarabaeidae). *Evolution* 50: 1219–1229.

Emlen, D. J. 1997. Alternative reproductive tactics and male-dimorphism in the horned beetle *Onthophagus acuminatus* (Coleoptera: Scarabaeidae). *Behavioral Ecology and Sociobiology* 41: 335–341.

Emlen, D. J. 1997. Diet alters male horn allometry in the beetle *Onthophagus acuminatus* (Coleoptera: Scarabaeidae). *Proceedings of the Royal Society London B* 264: 567–574.

Emlen, D. J. and Nijhout, H. F. 1999. Hormonal control of male horn length dimorphism in the dung beetle *Onthophagus taurus* (Coleoptera: Scarabaeidae). *Journal of Insect Physiology* 45: 45–53.

Enattah, N. S., Sahi, T., Savilhati, E., Terwilliger, J. D., Peltonen, L., and Jarvela, I. 2002. Identification of a variant associated with adult-type hypolactasia. *Nature Genetics* 30: 233–237.

Erwin, D. and Davidson, E. 2002. The Last Common Bilaterian Ancestor. *Development* 129: 3021–3032.

Extavour, C. G., Pang, K., Matus, D. Q., and Martindale, M. Q. 2005. *vasa* and *nanos* expression patterns in a sea anemone and the evolution of bilaterian germ cell specification mechanisms. *Evolution and Development* 7: 201–215.

Gardner, M. 1983. *Wheels, life, and other mathematical amusements*. New York: Freeman.

Falk, R. 2000. The Gene: A concept in tension. In: Beurton, P. J., Falk, R., and Rheinberger, H. J. (eds.) *The Concept of the Gene in Development and Evolution*. Cambridge: Cambridge University Press, pp. 317–348.

Farley, J. 1977. *The spontaneous generation controversy from Descartes to Oparin*. Baltimore: Johns Hopkins University Press.

Farmer, D. and Belin, A. 1992. Artificial life: the coming evolution. In: Langton, C., Taylor, C., Farmer, J. D., and Rasmussen, S., (eds.) *Artificial life II*. Redwood City, CA: Addison Wesley, pp. 815–840.

Feldman, M. W. and Laland, K. N. 1996. Gene-culture coevolutionary theory. *TREE* 11: 453–457.

Fisher, R. A. 1930. *The genetical theory of natural selection*. Oxford: Clarendon Press.

Finnerty, J. R., Pang, K., Burton, P., Paulson, D., and Martindale, M. Q. 2004. Origins of bilateral symmetry: Hox and dpp expression in a sea anemone. *Science* 304: 1335–1337.
Flack, J. C., Girvan, M., de Waal, F. B. M., and Krakauer, D. C. 2006. Policing stabilizes construction of social niches in primates. *Nature* 439: 426–429.
Fondon, J. W. 3rd and Garner, H. R. 2004. Molecular origins of rapid and continuous morphological evolution. *Proceedings of the National Academy of Sciences, USA* 101: 18058–18063.
Forrest, B. and Gross, P. G. 2004. *Creationism's trojan horse: the wedge of intelligent design*. Oxford: Oxford University Press.
Foucault, M. 1970. *The order of things: an archeology of the human sciences*. New York: Vintage, p. xv.
Frank, S. 1998. *Foundations of social evolution*. Princeton: Princeton University Press.
Frederickson, M. E., Greene, M. J., and Gordon, D. M. 2005. 'Devil's gardens' bedevilled by ants. *Nature* 437: 495–496.
Fry, I., 1995. Are the Different Hypotheses on the Emergence of Life as Different as they Seem? *Biology and Philosophy* 10: 389–417.
Fry, I. 2000. *The emergence of life on earth*. Rutgers University Press.
Fujisawa, T. 2006. Hydra is joining the bandwagon. *BioEssays* 28: 560–562.
Futuyma, D. J. 1998. *Evolutionary biology*. (3rd edn.). Sinauer.
Galton, F. 1877. Typical laws of heredity. *Proceedings of the Meetings of the Members of the Royal Institution of Great Britain* 8: 282–301.
Gayon, J. 1998. *Darwinism's struggle for survival: heredity and the hypothesis of natural selection*. Cambridge: Cambridge University Press.
Gerhart, J. C. and Kirschner, M. W. 1997. *Cells, embryos and evolution*. Malden, Massachusetts: Blackwell Science.
Gilbert, S. F. 1979. The metaphorical structuring of social perceptions. *Soundings* 62: 166–186.
Gilbert, S. F. 1997. Bodies of knowledge: Biology and the Intercultural University. In: Taylor, P. J., Halfon, S. E., and Edwards, P. N. (eds.) *Changing Life*. Minneapolis: Minnesota University Press. pp. 36–55.
Gilbert, S. F. 2000. Genes classical and genes developmental: The different use of genes in evolutionary synthesis. In: Beurton, P. J., Falk, R., and Rheinberger, H. J. (eds.) *The Concept of the Gene in Development and Evolution*. Cambridge: Cambridge University Press, pp. 178–192.
Gilbert, S. F. 2001. Ecological developmental biology: Developmental biology meets the real world. *Developmental Biology* 233: 1–12.
Gilbert, S. F. 2003c. *Developmental biology*. (7th edn.). Sinauer: Sunderland.
Gilbert, S. F. 2004. Mechanisms for the environmental regulation of gene expression. *Birth Defects Research: Embryo* 72: 291–299.
Gilbert S. F. 2006a. The generation of novelty: The province of developmental biology. *Biological Theory* 1(2): 209–212.
Gilbert, S. F. 2006b. *Developmental Biology*. (8th edn.). Sunderland, MA: Sinauer Associates, Inc. Publishers.
Gilbert, S. F. and Epel, D. 2009. *Ecological Developmental Biology*. Sinauer Associates, Sunderland, MA.
Gilbert, S. F., Opitz, J. M., and Raff, R. A. 1996. Resynthesizing evolutionary and developmental biology. *Developmental Biology* 173: 357–72.
Godfrey-Smith, P. 1996. *Complexity and the function of mind in nature*. Cambridge: Cambridge University Press.
Godfrey-Smith, P. 1998. Maternal Effects: On Dennett and *Darwin's Dangerous Idea*. *Philosophy of Science* 65: 709–720.
Godfrey-Smith, P. 1999. Genes do not Encode Information for Phenotypic Traits. In: Hitchcock, C. (ed.) *Current Debates in Philosophy of Science*. Malden, MA: Blackwell Publishing, pp. 275–289.
Godfrey-Smith, P. 2000a. Information, arbitrariness and selection: Comments on Maynard Smith. *Philosophy of Science* 67: 202–207.

Godfrey-Smith, P. 2000b. On the Theoretical Role of 'Genetic Coding'. *Philosophy of Science* 67: 26–44.
Godfrey-Smith, P. 2001a. Three Kinds of Adaptationism. In: Orzack, S. H. and Sober, E. (eds.) *Adaptationism and Optimality*. Cambridge: Cambridge University Press, pp. 335–357.
Godfrey-Smith, P. 2001b. Information and the argument from design. In: Pennock, R. T. (ed.) *Intelligent Design Creationism and its Critics: Philosophical, Theological, and Scientific Perspectives*. Cambridge, MA: Bradford/MIT Press, pp. 575–596.
Godfrey-Smith, P. 2003. *Theory and reality: an introduction to the philosophy of science*. Chicago: University of Chicago.
Godfrey-Smith, P. 2006b. The Strategy of Model-Based Science. *Biology and Philosophy* 21(5): 725–740.
Godfrey-Smith, P. (2007). Information in biology. In: Hull, D. and Ruse, M. (eds.) *The Cambridge Companion to the Philosophy of Biology*. Cambridge: Cambridge University Press.
Goodwin, B. 2001. *How the leopard changed its spots, second edition*. Princeton: Princeton University Press.
Goldschmidt, R. B. 1940. *The material basis of evolution*. New Haven: Yale University Press.
Gould, S. J. 1977. *Ontogeny and phylogeny*. Cambridge, MA: Harvard University Press.
Gould, S. J. 1989. *Wonderful life: the Burgess shale and the nature of history*. New York: Norton.
Gould, S. J. 1996. *Full house: the spread of excellence from Plato to Darwin*. New York: Random House.
Gould, S. J. and Eldredge, N. 1977. Punctuated equilibria: the tempo and mode of evolution reconsidered. *Paleobiology* 3: 115–151.
Gould, S. J., and Lewontin, R. C. 1979. The spandrels of San Marco and the Panglossian paradigm: a critique of the adaptationist program. *Proceedings of the Royal Society of London, Series B: Biological Sciences* 205: 581–98.
Grassie, W. 2005. Science, semiotics, and the sacred. In: Harper, Jr. C. L. (ed.) *Spiritual Information: 100 Perspectives on Science and Religion*. Philadelphia, PA: Templeton Foundation Press, p. 40.
Gray, A. 1881. *Structural botany*. (6th edn.). London: Macmillan.
Griesemer, J. and Wimsatt, W. 1989. Picturing Weismannism: A case study of conceptual evolution. In: Ruse, M. (ed.) *What the Philosophy of Biology is: Essays for David Hull*. Dordrecht, Netherlands: Kluwer Academic, pp. 75–137.
Griesemer, J. R. 2005. The informational gene and the substantial body: On the generalization of evolutionary theory by abstraction. In: Jones, M. R. and Cartwright, N. (eds.) *Idealization XII: Correcting the Model, Idealization and Abstraction in the Sciences*. Poznan Studies in the Philosophy of the Sciences and the Humanities, vol. 86. Amsterdam: Rodopi, pp. 59–115.
Griffiths, P. E. 2001. Genetic information: A metaphor in search of a theory. *Philosophy of Science* 68: 394–412.
Griffiths, P. E. 2002. The philosophy of molecular and developmental biology. In: Machamer, P. K. and Silberstein, M. (eds.) *Blackwell's Guide to the Philosophy of Science*. Malden, MA, Oxford: Blackwell Publishing.
Griffiths, P. E. 2007. Gene. In: Ruse, M. and Hull, D. (eds.) *The Cambridge Companion to Philosophy of Biology*. Cambridge: Cambridge University Press, 2007.
Griffiths, P. E. and Gray, R. D. 2004. The Developmental Systems Perspective: Organism-Environment Systems as Units of Development and Evolution. In: Pigliucci, M. and Preston, K. (eds.) *Phenotypic Integration: Studying the Ecology and Evolution of Complex Phenotypes*. Oxford University Press, pp. 390–408.
Griffiths, P. E. and Neumann-Held, E. 1999. The many faces of the gene. *BioScience* 49(8): 656–662.
Guido, N. J. et al. 2006. A bottom-up approach to gene regulation. *Nature* 439: 856–860.
Gurney, W. S. C. and Lawton, J. H. 1996. The population dynamics of ecosystem engineers. *Oikos* 76: 273–283.
Haig, D. 1997. The Social Gene. In: Krebs, J. and Davies, N. (eds.) *Behavioural Ecology*. (4th edn.). Oxford: Blackwell, pp. 284–304.

Haldane, J. B. S. 1929. The origin of life. *Rationalist Annual* 3: 1–10.

Haldane, J. B. S. 1949. *What is Life?* London: Lindsay Drummond.

Hammond, D. 2003. *The science of synthesis: Exploring the social implications of general systems theory*. Boulder, CO: University Press of Colorado.

Hanski, I. and Singer, M. C. 2001. Extinction-colonization and host-plant choice in butterfly metapopulations. *American Naturalist* 158: 341–353.

Haraway, D. 2003. *The Companion Species Manifesto: Dogs, People, and Significant Otherness*. Chicago: Prickly Paradigm Press.

Haught, J. F. 2000. *God after Darwin: A theology of evolution*. Boulder, CO: Westview Press.

Haught, J. F. 2003. *Deeper than Darwin: The prospect for religion in the age of evolution*. Boulder, CO: Westview Press.

Haworth, K. E. et al. 2001. Canine TCOF1: Cloning, chromosome assignment, and genetic analysis in dogs with different head types. *Mammalian Genome* 12: 622–629.

Hayes, T., Haston, K., Tsui, M., Hoang, A., Haeffele, C., and Vonk, A. 2003. Atrazine-induced hermaphroditism at 0.1 ppb in American leopard frogs (*Rana pipiens*): laboratory and field evidence. *Environ Health Perspect* 111: 568–575.

Hayles, N. K. 1999. *How we became posthuman: Virtual bodies in cybernetics, literature, and informatics*. Chicago: University of Chicago Press.

Hazen, R. M. 2005. *Genesis: The scientific quest for life's origin*. Washington: Joseph Henry Press.

Heims, S. J. 1991. *The cybernetics group*. Cambridge, MA: MIT Press.

Hofbauer, J. and Sigmund, K. 1998. *Evolutionary games and population dynamics*. Cambridge: Cambridge University Press.

Holland, J. H. 1992. *Adaptation in natural and artificial systems: an introductory analysis with applications to biology, control, and artificial intelligence*. (2nd edn.). Cambridge: MIT Press.

Holland, J. H. 1995. *Hidden order: how adaptation builds complexity*. Reading, MA: Helix Books.

Holland, L. Z. 2002. Heads or tails? Amphioxus and the evolution of anterior-posterior patterning in deuterostomes. *Developmental Biology* 241: 209–228.

Holland, L. Z., Laudet, V., Schubert, M. 2004. The chordate amphioxus: an emerging model organism for developmental biology.*Cellular and Molecular Life Sciences* 61:2290–2308.

Holt, R. D. 2005. On the integration of community ecology and evolutionary biology: Historical perspectives, and current prospects. In: Cuddington, K. and Beisner, B. (eds.) *Ecological paradigms lost: Routes of theory change*. New York: Academic Press.

Hooper, L. V., Wong, M. H., Thelin, A., Hansson, L., Falk, P. G., and Gordon, J. I. 2001. Molecular analysis of commensal host-microbial relationships in the intestine. *Science* 291: 881–884.

Hooper, L. V., Stappenbeck, T. S., Hong, C. V., and Gordon, J. I. 2003. Angiogenins: a new class of microcidal proteins involved in innate immunity. *Nature Immunology* 4: 269–273.

Hraber, P. T., Jones, T., and Forrest, S. 1997. The ecology of Echo. *Artificial Life* 3: 165–190.

Hull, D., 1974. *The Philosophy of Biological Science*. Englewood Cliffs, NJ: Prentice-Hall.

Hunt, J. and Simmons, L. W. 1998. Patterns of parental provisioning covary with male morphology in a horned beetle (*Onthophagus taurus*) (Coleoptera: Scarabaeidae). *Behavioral Ecology and Sociobiology* 42: 447–451.

Irwin, J. T., Bennett, V. A., and Lee, R. E. Jr. 2001. Diapause development in frozen larvae of the goldenrod gall fly, *Eurosta solidaginis* fitch (diptera: tephritidae). *The Journal of Comparative Physiology B* 171: 181–188.

Jablonka, E. and Lamb, M. J. 1995. *Epigentic inheritance and evolution*. Oxford: Oxford University Press.

Jablonka, E. and Lamb, M. 2005. *Evolution in four dimensions*. Cambridge, MA: MIT Press.

Jacob, F. 1973. *The logic of life: A history of heredity*. (B. E. Spillman, trans.). New York: Pantheon.

Jacob, F. 1977. Evolution and tinkering. *Science* 196: 1161–1166.

Jaynes, E. T. 1996. The Gibbs paradox. In: Smith, C. R., Erickson, G. J., and Neudorfer, P. O. (eds.) *Maximum Entropy and Bayesian Methods*. Dodrecht: Kluwer Academic.

Jenner, R. A. 2006. Unburdening evo-devo: ancestral attractions, model organisms, and basal baloney. *Development Genes and Evolution* 216: 385–394.

Johnson, P. E. 1999. Reclaiming America for Christ Conference. How the evolution debate can be won. http://www.coralridge.org/specialdocs/evolutiondebate.asp
Johnson, P. E. 2001. Is genetic information irreducible? In: Pennock, R. T. (ed.) *Intelligent design creationism and its critics: Philosophical, theological, and scientific perspectives*. Cambridge, MA: Bradford/MIT Press, pp. 543–547. Orig 1996 *Biol & Phil*, 11(4): 535–538.
Jones and Lawton 1995. (eds.) Linking Species & Ecosystems. Chapman & Hall.
Jones, C. G., Lawton, J. H., and Shachak, M. 1994. Organisms as ecosystem engineers. *Oikos* 69: 373–386.
Jones, C. G., Lawton, J. H. and Shachak, M. 1997. Positive and negative effects of organisms as physical ecosystem engineers. *Ecology* 78: 1946–1957.
Judson, H. F. 1979. *The eighth day of creation: Makers of the revolution in biology*. New York: Simon & Schuster.
Kant, I. 1993. *Critique of Judgement*, in *Great Books* 39: 461, Chicago: Encyl. Brittanica.
Kauffman, S. 1993. *The origins of order*. New York: Oxford University Press.
Kauffman, S. A. 1995. *At home in the universe: The search for the laws of self-organization and complexity*. New York: Oxford University Press.
Kauffman, S. 2000. *Investigations*. New York: Oxford University Press.
Kay, L. E. 2000. *Who wrote the book of life? A history of the genetic code*. Stanford, CA: Stanford University Press.
Keller, E. F. 1990. Physics and the emergence of molecular biology: A history of cognitive and political synergy. *Journal of the History of Biology* 23: 389–410.
Keller, E. F. 2000. *The century of the gene*. Cambridge, MA: Harvard University Press.
Keller, E. F. 2005. Revisiting "Scale-Free" Networks. *BioEssays* 27(1): 1060–1068.
Keller, E. and Lloyd, E. 1992. *Keywords in evolutionary biology*. Cambridge, MA: Harvard University Press.
Kimura, M. 1968. Evolutionary Rate at the Molecular Level. *Nature* 217: 624–626.
Kirk, M. and Kirk, D. A. 2004. Exploring germ-soma differentiation in *Volvox*. *Journal of Bioscience* 29: 143–152.
Kirkipatrick, M. and Lande, R. 1989. The evolution of maternal charaters. *Evolution* 43: 485–503.
Kirschner, M. W. and Gerhart, J. C. 2005. *The Plausibility of Life*. New Haven: Yale University Press.
Kitano, H. 2004. Biological Robustness. *Nature Reviews Genetics* 5: 826–837.
Kitano, H. 2006. Self-Extending Symbiosis: A Mechanism for Increasing Robustness Through Evolution. *Biological Theory* 1(1): 61–66.
Kitcher, P. 1984. 1953 and All That: A Tale of Two Sciences. *Philosophical Review* 93: 335–373.
Kussell, E. and Leibler, S. 2005. Phenotypic diversity, population growth, and information in fluctuating environments. *Science* 309: 2075–2078.
Laland, K. N., Odling-Smee, F. J., and Feldman, M. W. 1996. On the Evolutionary Consequences of Niche Construction. *Journal of Evolutionary Biology* 9: 293–316.
Laland, K. N., Odling-Smee, F. J. and Feldman, M. W. 1999. Evolutionary Consequences of Niche Construction and their Implications for Ecology. *Proceedings of the National Academy of Sciences USA* 96: 10242–10247.
Laland, K. N., Odling-Smee, F. J., and Feldman, M. W. 2001. Cultural niche construction and human evolution. *Evolutionary Biology* 14: 22–23.
Laland, K. N., Odling-Smee, F. J., and Feldman, M. W. 2005. On the Breadth and Significance of Niche Construction: A Reply to Griffiths, Okasha and Sterelny. *Biology and Philosophy* 20(1): 37–55.
Laland, K. L., Odling-Smee, J. and Gilbert, S.F. 2008. Evo-devo and niche construction: building bridges. *Journal of Experimental Zoology* B. 510: 549–566.
Lederberg, J. 1966. Remarks. *Current Topics in Developmental Biology* 1: ix–xiii.
Lee, K. E. 1985. Earthworms: their ecology and relation with soil and land use. New York: Academic Press.
Lee, T. I. et al. 2002. Transcriptional regulatory networks in *Saccharomyces cerevisiae*. *Science* 298: 799–804.

Lenski, R. E., Offria, C., Pennock, R. T., and Adami, C. 2003. The evolutionary origin of complex features. *Nature* 423: 139–144.

Leroi, A. M. 2000. The Scale Independence of Evolution. *Evolution and Development* 2(2): 67–77.

Lewontin, R. C. 1970. The Units of Selection. *Annual Review of Ecology and Systematics* 1: 1–18.

Lewontin, R. 1974. The Analysis of Variance and the Analysis of Causes. *American Journal of Human Genetics* 26: 400–11.

Lewontin, R. C. 1982. Organism and Environment. In: Plotkin, H. C. (ed.) *Learning, Development and Culture*. New York:Wiley, pp. 151–170.

Lewontin, R. C. 1983a. Gene, organism, and environment. In: Bendall, D. S. (ed.) *Evolution from Molecules to Men*. Cambridge: Cambridge University Press.

Lewontin, R. C. 1983b. The Organism as the Subject and Object of Evolution. *Scientia* 118: 63–82.

Lewontin, R. C. 1985. The Organism as Subject and Object of Evolution. In: Lewontin, R. C. and Levins, R. (eds.) *The Dialectical Biologist*. Cambridge: Harvard University Press.

Lewontin, R. C. 2000a. *The Triple Helix*. Cambridge, MA: Harvard University Press.

Lewontin, R. C. 2000b. Foreword to the 2nd edn. of Oyama, S. *The Ontogeny of Information*.

Lewontin, R. C., Rose, S., and Kamin, L. J. 1984. *Not in our genes: biology, ideology and human nature*. New York: Pantheon Books.

Li, L., Alderson, D., Doyle, J. C., and Willinger, W. 2004. A first-principles approach to understanding the Internet's router-level topology. *Proc ACM SIGCOMM*, Portland, Oregon, Aug. 30–Sept 2.

Lindblad-Toh, K. et al. 2005. Genome sequence, comparative analysis and haplotype structure of the domestic dog. *Nature* 438: 803–819.

Lloyd, S. 2001. Measures of complexity: a nonexhaustive list. *IEEE Control Systems Magazine* 21: 7–8.

Loewenstein, W. R. 1999. *The touchstone of life: Molecular information, cell communication, and the foundations of life*. Oxford: Oxford University Press.

Luisi, P. L. 2006. *The emergence of life: From chemical origins to synthetic biology*. Cambridge: Cambridge University Press.

Lyytinen, A., Brakefield, P. M., Lindstrom, L., Mappes, J. 2004. Does predation maintain eyespot plasticity in *Bicyclus anynana*? *Proceedings of the Biological Sciences* 271: 279–283.

Magie, C. R., Pang, K., Martindale, M. Q. 2005. Genomic inventory and expression of Sox and Fox genes in the cnidarian *Nematostella vectensis*. *Development Genes and Evolution* 215: 618–630.

Maienschein, J. 1992. Gene: Historical Perspectives. In: Keller, E. F. and Lloyd, E. A. (eds.) pp. 122–127.

Maley, C. C. 1994. The computational completeness of Ray's Tierran assembly language. In: Langton, C. G. (ed.) *Artificial Life III*. Reading, MA: Addison-Wesley, pp. 152–159.

Martin, E. 1992. The End of the Body? *American Ethnologist* 19: 121–140.

Martindale, M. Q. 2005. The evolution of metazoan axial properties. *Nature Review Genetics* 6: 917–927.

Martindale, M. Q., Pang, K., and Finnerty, J. R. 2004. Investigating the origins of triploblasty: 'mesodermal' gene expression in a diploblastic animal, the sea anemone *Nematostella vectensis* (phylum, Cnidaria; class, Anthozoa). *Development* 131: 2463–2474.

Matus, D. Q., Thomsen, G. H., and Martindale, M. Q. 2006. Dorso/ventral genes are asymmetrically expressed and involved in germ-layer demarcation during cnidarian gastrulation. *Current Biology* 16: 499–505.

Maynard-Smith, J. 1989. Weismann and modern biology. In: Harvey, P. and Partridge, L. (eds.) *Oxford Surveys in Evolutionary Biology*. Oxford: Oxford University Press, vol. 6, pp. 1–13.

Maynard Smith, J. 2000. The concept of information in biology. *Philosophy of Science* 67: 177–194.

Maynard Smith, J, Burian, R., Kauffman, S., Alberch, P., Campbell, J., Goodwin, B., Lande, R., Raup, D., and Wolpert, L. 1985. Developmental constraints and evolution. *Quarterly Review of Biology* 60: 265–87.

Maynard Smith, J. and Szathmáry, E. 1995. *The major transitions in evolution*. New York: W. H. Freeman.

Mayr, E. 1961. Cause and effect in biology, *Science* 134: 1501–1506.

Mayr, E. 1982. *The growth of biological thought*. Cambridge, MA: Harvard University Press/Belknap Press.

Mayr, E. 2004. 80 Years of watching evolutionary scenery. *Science* 305: 46–47.

McFall-Ngai, M. J. 2002. Unseen forces: the influence of bacteria on animal development. *Developmental Biology* 242: 1–14.

McMullen, E. 1985. Galilean idealization. *Studies in the History and Philosophy of Science* 16: 247–273.

McShea, D. 1991. Complexity and evolution: what everybody knows. *Biology and Philosophy* 6: 303–324.

McShea, D. 1994. Mechanisms of large-scale evolutionary trends. *Evolution* 48: 1747–1763.

McShea, D. 1996. Metazoan complexity and evolution: is there a trend? *Evolution* 50: 477–492.

McShea, D. 2005. The evolution of complexity without natural selection, a possible large-scale trend of the fourth kind. *Paleobiology* 31(2, Supplement): 146–156.

Meinhardt, H. 2002. Radial-symmetric hydra and the evolution of the bilateral body plan: an old body became a young brain. *Bioessays* 24: 185–191.

Michod, R. 1999. *Darwinian dynamics: Evolutionary transitions in fitness and individuality*. Princeton: Princeton University Press.

Millikan, R. G. 1984. *Language, thought, and other biological categories: New foundations for realism*. Cambridge, MA: MIT Press.

Miner, B. G., Sultan, S. E., Morgan, S. G., Padilla, D. K., and Reyles, R. A. 2005. Ecological consequences of phenotypic plasticity. *TREE* 20: 685–691.

Mitchison, G. J. 1977. Phyllotaxis and the Fibonacci series. *Science* 196: 270–275.

Mitmann, G. and Fausto-Sterling, A. 1992. Whatever Happened to Planaria? CM Child and the Physiology of Inheritance. In: Clarke, A. and Fujimura, J. (eds.) *The Right Tool for the Job: At Work in 20th Century Life Sciences*. Princeton : Princeton University Press.

Moczek, A. P. 2006. Integrating micro and macroevolution of development through the study of horned beetle. *Heredity* 97: 168–178.

Monod, J. 1970. *Le hasard et la nécessité*. Paris: Seuil. English transl.: *Chance and Necessity: An Essay on the Natural Philosophy of Modern Biology*. New York: Vintage Books, 1971.

Monod, J. (1971). Chance and necessity. (A. Wainhouse, Trans.) New York: Knopf.

Moore, G. 2002. Nietsche, spencer, and the ethics of evolution. *Journal of Nietzsche Studies* 23: 1–20.

Moore, K. A. and Lemischka, I. R. 2006. Stem cells and their niches. *Science* 311: 1880–1885.

Morange, M. 1998. *A history of molecular biology*. (M. Cobb, trans.) Cambridge, MA: Harvard University Press.

Morange, M. 2005. *Les secrets du vivant.Contre la pensée unique en biologie*. Paris: La Découverte.

Moss, L. 2003. *What genes can't do*. Cambridge, MA: MIT/Bradford.

Müller-Lauter, W. 1978. Der Organismus als innerer Kampf: Der Einfluß von Wilhelm Roux auf Friedrich Nietzsche. *Nietzsche-Studien* 7: 189–223.

National Science Foundation. 2005. An Integrated Developmental Biology: Workshop Report. Booklet 08053.

Neff, M. W. and Rine, J. 2006. A fetching model system. *Cell* 124: 229–231.

Nelkin, D., and Lindee, M. S. 1995. *The DNA mystique: The gene as a cultural icon*. New York: W. H. Freeman.

Nijhout, H. F. 1990. Metaphors and the role of genes in development. *BioEssays* 12: 444–446.

Nijhout, H. F. 2003. Development and evolution of adaptive polyphenisms. *Evolution and Development* 5: 9–18.

Niklas, K. J. 1988. The role of phyllotactic pattern as a 'developmental constraint' on the interception of light by leaf surfaces. *Evolution* 42: 1–16.

Nitecki, M. (ed.) 1990. *Evolutionary Innovations*. Chicago: University of Chicago Press.

Nowak, M. and Sigmund, K. 2004. Evolutionary Dynamics of Biological Games. *Science* 303: 793–799.

Odling-Smee, F. J. 1988. Niche constructing phenotypes. In. Plotkin, H. C. (ed.) *The Role of Behavior in Evolution*. Cambridge: MIT Press, pp. 73–132.

Odling-Smee, F. J., Laland, K. N. et al. 1996. Niche Construction. *American Naturalist* 147: 641–648.

Odling-Smee, F. J., Laland, K. N., and Feldman, M. W. 2003. *Niche construction: the neglected process in evolution*. Princeton: Princeton University Press.

Odum, 1989. Ecology and our Endangered Life-Support Systems. Sinauer.

Okasha, S. 2006. *Evolution and the Levels of Selection*. Oxford: Oxford University Press.

O'Malley, M. A. and Dupré, J. 2005. Fundamental issues in systems biology. *BioEssays* 27: 1270–1276.

O'Neill, R. V., DeAngelis, D. L., Waide, J. B., and Allen, T. F. H. 1986. *A hierarchical concept of ecosystems*. Princeton: Princeton University Press.

Ostwald, W. 1917. *Introduction to theoretical and applied colloid science: The world of neglected dimension*. New York: M. H. Fischer.

Ozbudak, E. M. et al. 2004. Multistability in the lactose utilization network of *Escherichia coli*. *Nature* 427: 737–740.

Oyama, S. 2000a. *Evolution's eye: A systems view of the biology-culture divide*. Durham, NC: Duke University Press.

Oyama, S. 2000b. *The ontogeny of information: Developmental systems and evolution* (2nd edn., revised and expanded). Durham, NC: Duke University Press.

Oyama, S. 2001. Terms in Tension: What Do You Do When All the Good Words are Taken? In: Oyama, S., Griffiths, P. E., and Gray, R. D. (eds.) pp. 177–193.

Oyama, S. 2002. The nurturing of natures. In Grunwald, A., Gutmann, M., and Neumann-Held, E. M. (eds.) *On human nature: Anthropological, biological, and philosophical foundations*. Wissenschaftsethik und Technikfolgenbeurteilung Band 15. Berlin: Springer Verlag, pp. 163–170.

Oyama, S., Griffiths, P. E., and Gray, R. D. (eds.) 2001. *Cycles of contingency: Developmental systems and evolution*. Cambridge, MA: MIT Press/Bradford Books.

Packard, N. 1989. Intrinsic adaptation in a simple model for evolution. In Langton, C. (ed.) *Artificial life*. Redwood City, CA: Addison-Wesley, pp. 141–155.

Passamaneck, Y. J., and Di Gregorio, A. 2005. *Ciona intestinalis*: chordate development made simple. *Developmental Dynamics* 233: 1–19.

Pearson, H. 2006. What is a gene? *Nature* 441: 399–401.

Pennock, R. T. (ed.) 2001. *Intelligent design creationism and its critics: Philosophical, theological, and scientific perspectives*. Cambridge, MA: Bradford/MIT Press.

Peterson, K. and Davidson, E. 2000. Regulatory Evolution and the Origin of the Bilaterians. *Proceedings of the National Academy of Science* 97(9): 4430–4433.

Pigliucci, M. and Preston, K. (eds.) 2004. *Phenotypic integration. Studying the ecology and evolution of complex phenotypes*. Oxford: Oxford University Press.

Pittendrigh, C. 1958. Adaptation, natural selection, and behavior. In: Roe, A. and Simpson, G. G. (eds.) *Behavior and evolution*. New Haven, CT: Yale University Press, pp. 390–416.

Poccia, D. 2006. Beyond the model organism. *The Journal of Experimental Zoology A* 305: 537.

Polkinghorne, J. 2000. *Faith, science and understanding*. New Haven, CT: Yale University Press.

Polkinghorne, J. 2003. *Belief in God in an age of science.The Terry Lectures*. New Haven, CT: Yale University Press.

Poundstone, W. 1985. *The recursive universe*. Chicago: Contemporary Books.

Price, G. 1972. Extension of Covariance Selection Mathematics. *Annuals of Human Genetics* 35: 485–90.

Provine, W. 1971. *On the Origins of Theoretical Population Genetics*. Chicago: Chicago University Press.

Raff, R. 1996. *The shape of life: Genes, development and the evolution of animal form*. Chicago, Chicago University Press.

Raff, R. and Kauffman, T. 1983. *Embryos, genes and evolution: The developmental-genetic basis of evolutionary change*. New York: Macmillan; London: Collier Macmillan.

Raser, J. M. and O'Shea, E. K. 2005. Noise in gene expression: origins, consequences, and control. *Science* 309: 2010–2013.

Rasmussen, N. 1987. A new model of developmental constraints as applied to the *Drosophila* system. *Journal of Theoretical Biology* 127: 271–301.

Ray, T. S. 1992. An approach to the synthesis of life. In: Langton, C. Taylor, C., Farmer, D., and Rasmussen, S. (eds.) *Artificial life II*. Redwood City: Addison-Wesley, pp. 371–408.

Reeve, H. K. and Sherman, P. W. 1993. Adaptation and the goals of evolutionary research. *Quarterly Review of Biology* 68: 1–32.

Relyea, R. A. and Mills, N. 2003. Predator-induced stress makes the pesticide carbaryl more deadly to gray treefrog tadpoles (*Hyla versicolor*).*Proceedings of the National Academy of the Sciences, USA* 98: 2491–2496.

Rendell, P. 2001. A Turing machine in Conway's Game of Life. Available on the web at www.cs.ualberta.ca/~bulitko/F02/papers/tm_words.pdf. Accessed November 2006.

Rhee, K.-J., Sethupathi, P. Driks, A. Lanning, D. K., and Knight, K. L. 2004. Role of commensal bacteria in development of gut-associated lymphoid tissues and the pre-immune repertoire. *Journal of Immunology* 172: 1118–1124.

Rheinberger, H. J. and Müller-Wille, S. 2004. Gene. *Stanford Encyclopedia of Philosophy*. http://www.seop.leeds.ac.uk/entries/gene/

Richerson, P. J. and Boyd, R. 2005. *Not by genes alone. How culture transformed human evolution*. Chicago: The University of Chicago Press.

Ridley, M. 2000. *Mendel's demon: Gene justice and the complexity of life*. London: Weidenfeld and Nicholson.

Robert, J. S. 2004. *Embryology, epigenesis and evolution: Taking development seriously*. Cambridge: Cambridge University Press.

Rolston, H., III. 1999. *Genes, genesis and God: Values and their origins in natural and human history*. Gifford Lectures, University of Edinburgh, 1997–1998. Cambridge: Cambridge University Press.

Rosenberg, A. 1985. *The structure of biological science*. Cambridge: Cambridge University Press.

Rosenberg, A. 1994. *Instrumental biology or the disunity of science*. Chicago: The University of Chicago Press.

Roth, G., Blanke, J., and Wake, D. B. 1994. Cell size predicts morphological complexity in the brains of frogs and salamanders. *Proceedings of the National Academy of the Sciences, USA* 91: 4796–800.

Roughgarden, J. 1979. *Theory of populations genetics and evolutionary ecology: An introduction*. Upper Saddle River: Prentice Hall.

Roux, 1881. *Der kampf der theile im organismus*. Leipzig: Engelmann.

Ruse, M. 2003. *Darwin and Design: Does Evolution have a Purpose?* Cambridge, MA: Harvard University Press.

Ruse, M. 2006a. Forty Years a Philosopher of Biology: Why EvoDevo Makes Me Still Excited About My Subject. *Biological Theory* 1(1): 35–37.

Ruse, M. 2006b. *Darwinism and its Discontents*. Cambridge: Cambridge University Press.

Ruse, M., 2006c. Bare-Knuckle Fighting: EvoDevo versus Natural Selection. *Biological Theory* 1(4): 402–403.

Sampson, J. R. 1984. *Biological information processing*. New York: John Wiley & Sons.

Sarkar, S. 1996a. Biological information: A skeptical look at some central dogmas of molecular biology. In: Sarkar, S. (ed.) *The philosophy and history of molecular biology: New perspectives*. Dordrecht, Netherlands: Kluwer Academic, pp. 187–231.

Sarkar, S. 1996b. Decoding "coding"—information and DNA. *BioScience* 46: 857–864.

Savolainen, P., Zhang, Y. P., Luo, J., Lundeberg, J., and Leitner, T. 2002. Genetic evidence for an East Asian origin of domestic dogs. *Science* 298: 1610–1613.

Satchell, J. E. (ed.) 1983. *Earthworm ecology: from Darwin to Vermiculite*. New York: Chapman and Hall.

Scadden, D. T. 2006. The stem-cell niche as an entity of action. *Nature*. 441. 1075–1079.

Schlichting, C. 2003. Origins of differentiation via phenotypic plasticity. *Evolution and Development* 5(1): 98–105.

Schlichting, C. D. and Pigliucci, M. 1998. *Phenotypic evolution: a reaction norm perspective*. Sunderland, MA : Sinauer.

Schwenk, K. and Wagner, G. P. 2004. The Relativism of Constraints on Phenotypic Evolution. In: Pigliucci, M. and Preston, K. (eds.) *Phenotypic Integration: Studying the Ecology and Evolution of Complex Phenotypes*. Oxford University Press, pp. 390–408.

Schwilk, D. W. 2003. Flammability is a niche construction trait: canopy architecture affects fire intensity. *American Naturalist* 162: 725–733.

Schilling, T. F. and Webb, J. 2007. J Exp Zoolog B Mol Dev Evol. 308(5): 515–522.

Seiglie, M. 2005. A *Good News* interview with Michael Behe, Ph.D. *The Good News*, May-June. http://www.gnmagazine.org/issues/gn58/dna.htm

Shostak, S. 1998. *Death of life: a legacy of molecular biology*. London: Macmillan.

Simon, Herbert, 1996/1962. *The sciences of the artificial*. (3rd edn.). Cambridge MA: MIT Press.

Simpson, G. G. 1949. *The meaning of evolution*. New Haven, CT: Yale University Press.

Smith, B. H. (in press). *Natural reflections: Human cognition at the nexus of science and religion*. Terry Lectures, New Haven, CT: Yale University.

Smith, J. E. H. 2005. Degeneration and hybridism in the early modern species debate: Towards the philosophical roots of the creation-evolution controversy. In: Wolfe, C. T. (ed.) *Monsters & Philosophy. Texts in Philosophy*. vol. 3. London: College Publications, pp. 109–130.

Smith, R. H. and Bedau, M. A. 2000. Is ECHO a complex adaptive system? *Evolutionary Computation* 8: 419–442.

Solomon, E. P., Berg, L. R., Martin, D. W., and Villee, C. 1996. *Biology* (4th edn.). Fort Worth: Saunders College Publishing.

Spradling, A., Drummond-Barbosa, D., and Kai, T. 2001. Stem cells find their niche. *Nature* 414: 98–104.

Stappenbeck, T. S., Hooper, L. V., and Gordon, J. I. 2002. Developmental regulation of intestinal angiogenesis by indigenous microbes via Paneth cells. *Proceedings of the National Academy of the Sciences, USA* 99: 15451–15455.

Sterelny, K. 2001. Niche construction and the extended replicator. In: Oyama, S., Griffiths, P. E., and Gray, R. D. (eds.) *Cycles of Contingency: Developmental Systems and Evolution*. Cambridge, MA: MIT Press, pp. 333–349.

Sterelny, K. 2004. Symbiosis, evolvability and modularity. In: Schlosser, G. and Wagner, G. (eds.) *Modularity in Development and Evolution*. Chicago: University of Chicago Press, pp. 490–516.

Sterelny, K. 2005. Made by each other: Organisms and their Environment. *Biology and Philosophy* 20(1): 21–36.

Sterelny, K. and Griffiths, P. E. 1999. *Sex and death: an introduction to the philosophy of biology*. Chicago: University of Chicago Press.

Sterelny, K., Smith, K. C., and Dickison, M. 1996. The extended replicator. *Biology and Philosophy* 11: 377–403.

Stern, C. D. 2005. The chick: A great model system becomes even greater. *Developmental Cell* 8: 9–17.

Stockard. C. R. 1941. *The genetic and endocrine basis for differences in form and behavior*. Philadelphia: Wistar Institute Press.

Stotz, K. C. and Griffiths, P. E. 2001. Dancing in the dark: Evolutionary psychology and the argument from design. In: Scher, S. and Rauscher, M. (eds.) *Evolutionary psychology: Alternative approaches*. Dordrecht: Kluwer, pp. 135–160.

Stotz, K. and Griffiths, P. E. 2004. Genes: Philosophical analyses to put to the test. *History and Philosophy of the Life Sciences* 26: 5–28.

Sultan, S. E. 2003. Phenotypic plasticity in plants: a case study in ecological development.*Evolution and Development* 5: 25–33.

Suppes, P. 1960. A Comparison of the Meaning and Uses of Models in Mathematics and the Empirical Sciences. *Synthese* 12: 287–301.

Sutter, N. B. and Ostrander, E. A. 2004. Dog star rising: The canine genetic system. *Nature Reviews Genetics* 5: 900–910.

Szathmáry, E. 2002. The gospel of inevitability. *Nature* 419(6909): 779.

Szilard, L. 1929/1964. On the decrease of entropy in a thermodynamic system by the intervention of intelligent beings. Anatol Rapoport & Mechthilde Knodler, trans. *Behavioral Science* 9: 301–310.

Tessmar-Raible, K. and Arendt, D. 2003. Emerging systems: between vertebrates and arthropods. *Current Opinion in Genetics and Development* 13: 331–340.

Thomson, J. N. 1994. *The coevolutionary process*. Chicago: The University of Chicago Press.

Thomson-Jones, M. 2005. Idealization and Abstraction: A Framework. In: Thomson-Jones, M. and Cartwright, N. (eds.) *Idealization XII: Correcting the Model*. Amsterdam: Rodopi, pp. 173–217.

Tooby, J. and Cosmides, L. 1992. The psychological foundations of culture. In: Barkow, J. H., Cosmides, L., and Tooby, J. (eds.) *The adapted mind*. Oxford: Oxford University Press, pp. 19–136.

Tumpey, T. M. et al. 2005. Characterization of the reconstructed 1918 Spanish influenza pandemic virus. *Science* 310: 77–80.

Turner, J. S. 2000. *The extended organism: the physiology of animal-built structures*. Cambridge, MA: Harvard University Press.

van der Weele, C. 1999. *Images of Development: Environmental Causes in Ontogeny*. Albany: State University of New York Press.

Van Straalen, N. M. and Roelofs, D. 2006. *An introduction to ecological genomics*. Oxford: Oxford University Press.

Van Till, H. J., and Johnson, P. E. 1993. God and evolution: an exchange. *First Things, 34* (June/July): 32–41. http://www.firstthings.com/ftissues/ft9306/johnson.html

Vogel, S. 1988. *Life's Devices: The Physical World of Animals and Plants*. Princeton, NJ: Princeton University Press.

Voigt, B. F., Kudaravalli, S., Wen, X., and Pritchard, J. K. 2006. A map of recent positive selection in the human genome. *PLOS Biology* 0446–0458.

Wada, H. 2001. Origin and evolution of the neural crest: a hypothetical reconstruction of its evolutionary history. *Development, Growth and Differentiation* 43: 509–520.

Waddington, C. H. 1942. L'épigénotype. *Endeavour* 1: 18–20.

Waddington, C. H. 1947. *Organisers and Genes*. Cambridge: Cambridge University Press.

Waddington, C. H. 1953. Epigenetics and evolution. In: Brown, R. and Danielli, J. F. (eds.) *Evolution*. (SEB Symposium VII). Cambridge: Cambridge University Press, pp. 186–199.

Waddington, C. H. 1959a. Evolutionary adaptation. In: Tax, S. (ed.) *The evolution of life. Evolution after Darwin*. vol. 1, 381–402.

Waddington, C. H. 1959b. Canalisation of Development and the Inheritance of Acquired Characters. *Nature* 183: 1654–1655.

Waddington, C. H. 1969. Paradigm for an evolutionary process. In: Waddington, C. H. (ed.) *Towards a theoretical biology*. Edinburgh, UK: Edinburgh University Press, pp. 106–128.

Wang, E. T., Kodoma, G., Baldi, P., and Moyzis, R. K. 2006. Global landscape of recent inferred Darwinian selection for *Homo sapiens*. *Proceedings of the National Academy of Sciences, USA* 103: 135–140.

Walt, S. M. 1998. International relations: One world, many theories. *Foreign Policy* 110: 29–45.

Waters, K. 1990. Why the Anti-Reductionist Consensus Won't Survive the Case of Classical Mendelian Genetics. *PSA 1990*: vol. 1 East Lansing: Philosophy of Science Association, pp. 125–139.

Waters, K. 1994. Genes made molecular. *Philosophy of Science* 61: 163–185.

Watson, R. A., 2006. *Compositional evolution*. Cambridge, MA: MIT Press.

Webb, J. and Schilling, T. 2006. Zebrafish in comparative context: A symposium. *Integrative Computational Biology* in press.
Weber, M. 2005. *Philosophy of Experimental Biology*. Cambridge: Cambridge University Press.
Weber, B. H. and Depew, D. J. 1996. Natural Selection and Self-Organization. *Biology and Philosophy* 11(1): 33–65.
Weisberg, M. 2007. Who is a Modeler? To appear in the *British Journal for the Philosophy of Science* 58(2): 207–233.
Wells, J. 2006. *The politically incorrect guide to Darwinism and intelligent design*. Washington, DC: Regnery Publishing.
West-Eberhard, M. J. 2003. *Developmental plasticity and evolution*. Oxford: Oxford University Press.
Wilkins, A, S. 2002. *The evolution of developmental pathways*. Sunderland, MA: Sinauer Associates, p. 520.
Williams, G. C. 1966. *Adaptation and natural selection*. Princeton, NJ: Princeton University Press.
Williams, G. C. 1992a. *Natural selection: Domains, levels, and challenges*. Oxford: Oxford University Press.
Williams, G. C. 1992b. Gaia, nature worship, and biocentric fallacies. *The Quarterly Review of Biology* 67: 479–486.
Wimsatt, W. 1986. Developmental constraints, generative entrenchment, and the innate-acquired distinction. In: Bechtel, W. (ed.) *Integrating scientific disciplines*. The Netherlands: Martinus-Nijhoff, pp. 185–208.
Wimsatt, W. C. 1997. Aggregativity: Reductive Heuristics for Finding Emergence. *Philosophy of Science* 64(Suppl. 2): S372–S384.
Wimsatt, W. 2001. Generative Entrenchment and the Developmental Systems Approach to Evolutionary Processes. In Oyama, S., Griffiths, P. E., and Gray, R. D. (eds), *Cycles of Contingency: Developmental Systems and Evolution*. Cambridge: MIT Press, pp. 219–237.
Wimsatt, W. and Shank, J. 1988. Two constraints on the evolution of complex adapations and the means for their avoidance. In Nitecki, M. and Nitecki, D. (eds.) *Evolutionary Progress*. Chicago: University of Chicago Press, pp. 231–273.
Winther, R. (2006). On the dangers of making abstract scientific constructs ontologically independent: Taking Richard Levins' warnings seriously. *Biology and Philosophy* 21: 703–724.
Woese, C. 2002. On the evolution of cells. *PNAS* 99(13): 8742–7.
Wolf, J. B. 2000. Indirect genetic effects and gene interactions. In: Wolf, J. B., Brodie III, E. D., and Wade, M. J.(eds.) *Epistasis and the Evolutionary Process*. Oxford: Oxford University Press.
Wolfe, C. T. (ed.) 2005. *Monsters & philosophy*. London: College Publications.
Wolfram, S. 1994. *Cellular automata and complexity*. Reading, MA: Addison-Wesley.
Wolfram, S. 2002. *A new kind of science*. Champaign, IL: Wolfram Media.
Wright, J. P. and Jones, C. G. 2006. The concept of organisms as ecosystem engineers ten years on: progress, limitations and challenges. *BioScience* 3: 203–209.
Young, R. M. 1985. *Darwin's metaphor: Nature's place in Victorian culture*. Cambridge: Cambridge University Press.
Zakany, J. and Duboule, D. 1999. Hox Genes on Digit Development and Evolution. *Cell Tissue Research* 296: 19–25.

Index

MIX
Papier aus verantwortungsvollen Quellen
Paper from responsible sources
FSC® C105338

If you have any concerns about our products,
you can contact us on
ProductSafety@springernature.com

In case Publisher is established outside the EU,
the EU authorized representative is:
Springer Nature Customer Service Center GmbH
Europaplatz 3, 69115 Heidelberg, Germany

Printed by Libri Plureos GmbH
in Hamburg, Germany